Lecture Notes in Computer Scienc

T0238543

Commenced Publication in 1973
Founding and Former Series Editors:
Gerhard Goos, Juris Hartmanis, and Jan van Leeuwen

Marina L. Gavrilova   C.J. Kenneth Tan
Yingxu Wang   Keith C.C. Chan (Eds.)

Transactions on Computational Science V

Special Issue on Cognitive Knowledge Representation

 Springer

Editors-in-Chief

Marina L. Gavrilova
University of Calgary, Department of Computer Science
2500 University Drive N.W., Calgary, AB, T2N 1N4, Canada
E-mail: marina@cpsc.ucalgary.ca

C.J. Kenneth Tan
Exascala Ltd.
Unit 9, 97 Rickman Drive, Birmingham B15 2AL, UK
E-mail: cjtan@exascala.com

Guest Editors

Yingxu Wang
University of Calgary
Theoretical and Empirical Software Engineering Research Centre (TESERC)
2500 University Drive N.W., Calgary, AB, T2N 1N4, Canada
E-mail: yingxu@ucalgary.ca

Keith C.C. Chan
The Hong Kong Polytechnic University, Department of Computing
Hung Hom, Kowloon, Hong Kong
E-mail: cskcchan@inet.polyu.edu.hk

Library of Congress Control Number: Applied for

CR Subject Classification (1998): I.2.4, I.2.6, H.2.8, D.2.4, F.4

ISSN 0302-9743 (Lecture Notes in Computer Science)
ISSN 1866-4733 (Transaction on Computational Science)

ISBN 978-3-642-02096-4 Springer Berlin Heidelberg New York

springer.com

© Springer-Verlag Berlin Heidelberg 2009

Typesetting: Camera-ready by author, data conversion by Scientific Publishing Services, Chennai, India
Printed on acid-free paper   SPIN: 12669981   06/3180   5 4 3 2 1 0

LNCS Transactions on Computational Science

Computational science, an emerging and increasingly vital field, is now widely recognized as an integral part of scientific and technical investigations, affecting researchers and practitioners in areas ranging from aerospace and automotive research to biochemistry, electronics, geosciences, mathematics, and physics. Computer systems research and the exploitation of applied research naturally complement each other. The increased complexity of many challenges in computational science demands the use of supercomputing, parallel processing, sophisticated algorithms, and advanced system software and architecture. It is therefore invaluable to have input by systems research experts in applied computational science research.

Transactions on Computational Science focuses on original high-quality research in the realm of computational science in parallel and distributed environments, also encompassing the underlying theoretical foundations and the applications of large-scale computation. The journal offers practitioners and researchers the opportunity to share computational techniques and solutions in this area, to identify new issues, and to shape future directions for research, and it enables industrial users to apply leading-edge, large-scale, high-performance computational methods.

In addition to addressing various research and application issues, the journal aims to present material that is validated – crucial to the application and advancement of the research conducted in academic and industrial settings. In this spirit, the journal focuses on publications that present results and computational techniques that are verifiable.

Scope

The scope of the journal includes, but is not limited to, the following computational methods and applications:

- Aeronautics and Aerospace
- Astrophysics
- Bioinformatics
- Climate and Weather Modeling
- Communication and Data Networks
- Compilers and Operating Systems
- Computer Graphics
- Computational Biology
- Computational Chemistry
- Computational Finance and Econometrics
- Computational Fluid Dynamics
- Computational Geometry

- Computational Number Theory
- Computational Physics
- Data Storage and Information Retrieval
- Data Mining and Data Warehousing
- Grid Computing
- Hardware/Software Co-design
- High-Energy Physics
- High-Performance Computing
- Numerical and Scientific Computing
- Parallel and Distributed Computing
- Reconfigurable Hardware
- Scientific Visualization
- Supercomputing
- System-on-Chip Design and Engineering

Editorial

It is recognized that all forms of intelligence are memory-based. Therefore, the mechanisms of knowledge representation in memory play a crucial rule in cognitive informatics, computational science, intelligence science, and knowledge engineering. Although conventional technologies dealt with external knowledge representation in computing and machines, this special issue on cognitive knowledge representation puts emphasis on internal knowledge representation mechanisms of the brain and their engineering applications. The latest research results of internal knowledge representation at the logical, functional, physiological, and biological levels are explored from multiple disciplines in order to put together a complete picture to explain the mechanisms and methodologies of internal knowledge representations of the brain and their impact on computing, artificial intelligence, and computational intelligence.

Cognitive knowledge representations investigate novel internal knowledge representation theories, methodologies, mathematical means, models, algorithms, technological implementations, and computational simulations. The structure of cognitive knowledge representation encompasses the following areas:

(I) Internal Knowledge Representation Theories
- Nature of knowledge as a result of learning
- Concept algebra
- The object-attribute-relation (OAR) model
- Concept networks
- Semantic networks
- Relationship between internal/external knowledge
- Mathematical means for modeling knowledge

(II) Logical Models of Cognitive Knowledge
- Mathematical models of knowledge
- Logic-based modeling of knowledge
- Process-based modeling of knowledge
- Rule-based modeling of knowledge
- Knowledge bases
- Cognitive complexity of knowledge
- Formal inferences based on knowledge
- Fuzzy logical representation of knowledge

(III) Neuroinformatic Models of Cognitive Knowledge
- Transformation between data and knowledge
- Roles of intelligence in knowledge generation
- Cognitive informatics
- Knowledge representation at the neural level
- Knowledge representation at the cognitive level
- Knowledge representation and linguistics
- Knowledge and memory

(IV) Knowledge Manipulation Technologies
- Knowledge engineering methodologies
- Autonomous knowledge processing
- Intelligent data mining
- Knowledge modeling
- Knowledge acquisition
- Knowledge manipulation
- Knowledge transformation
- WordNet and ConceptNet

This special issue on *Cognitive Knowledge Representation* of the journal *Transactions on Computational Science* presents the latest advances in internal knowledge representation in the brain, autonomous machines, and computational intelligence. This special issue includes 12 selected and refined papers from the IEEE 7th/6th International Conference on Cognitive Informatics (ICCI 2008/2007) as well as new contributions as highlighted below.

Yingxu Wang presents "Toward a Formal Knowledge System Theory and Its Cognitive Informatics Foundations." Knowledge science and engineering are an emerging field that studies the nature of human knowledge and its manipulations such as acquisition, representation, creation, composition, memorization, retrieval, and depository. This paper presents the nature of human knowledge and its mathematical models, internal representations, and formal manipulations. The taxonomy of knowledge and the hierarchical abstraction model of knowledge are investigated. Based on a set of mathematical models of knowledge and the object-attribute-relation (OAR) model for internal knowledge representation, rigorous knowledge manipulations are formally described by concept algebra. A coherent framework of formalized knowledge systems is modeled based on the analyses of roles of formal and empirical knowledge. Subsequently the theory of knowledge acquisition and the cognitive model of knowledge spaces are systematically developed.

Du Zhang studies "On Temporal Properties of Knowledge Base Inconsistency." Inconsistent knowledge and information in the real world often have to do with not only what conflicting circumstances are but also when they happen. As a result, scrutinizing only the logical forms that cause inconsistency may not be adequate. This paper describes research work on the temporal characteristics of inconsistent knowledge that can exist in an intelligent system. The author provides a formal definition for temporal inconsistency that is based on a logical-conflicting and temporal-coinciding dichotomy. The interval temporal logic underpins the treatment of temporally inconsistent propositions in a knowledge base. The author also proposes a systematic approach to identifying conflicting intervals for temporally inconsistent propositions. The results help delineate the semantic difference between the classical and temporal inconsistency.

Douglas S. Greer invesgates "Images as Symbols: An Associative Neurotransmitter-Field Model of the Brodmann Areas." The ability to associate images is the basis for learning relationships involving vision, hearing, tactile sensation, and kinetic motion. A new architecture is described that has only local, recurrent connections, but can directly form global image associations. This architecture has many similarities to the structure of the cerebral cortex, including the division into Brodmann areas, the distinct internal and external lamina, and the pattern of neuron interconnection. The images are represented as neurotransmitter fields, which differ from neural fields in the underlying principle that the state variables are not the neuron action potentials, but the chemical concentration of neurotransmitters in the extracellular space. The neurotransmitter cloud hypothesis, which asserts that functions of space, time, and frequency are encoded by the density of identifiable molecules, allows the abstract mathematical power of cellular processing to be extended by incorporating a new chemical model of computation. This makes it possible for a small number of neurons, even a single neuron, to establish an association between arbitrary images. A single layer of neurons, in effect, performs the computation of a two-layer neural network. Analogous to the bits in an SR flip-flop, two arbitrary images can hold each other in place in an association processor and thereby form a short-term image memory. Just as the reciprocal voltage levels in a flip-flop can produce a dynamical system with two stable states, reciprocal-image pairs can generate stable attractors thereby allowing the images to serve as symbols. Spherically symmetric wavelets, identical to those found in the receptive fields of the retina, enable efficient image computations. Noise reduction in the continuous wavelet transform representations is possible using an orthogonal projection based on the reproducing kernel. Experimental results demonstrating stable reciprocal-image attractors are presented.

Jun Hu and Guoyin Wang present "Knowledge Reduction of Covering Approximation Space." Knowledge reduction is a key issue in data mining. In order to simplify the covering approximation space and mining rules from it, Zhu proposed a reduction of covering approximation space which does not rely on any priorly given concept or decision. Unfortunately, it could only reduce absolutely redundant knowledge. To reduce relatively redundant knowledge with respect to a given concept or decision, the problem of relative reduction is studied in this paper. The condition in which an element of a covering is relatively reducible is discussed. By deleting all relatively reducible elements of a covering approximation space, one can get the relative reduction of the original covering approximation space. Moreover, one can find that the covering lower and upper approximations in the reduced space are the same as in the original covering space. That is, it does not decrease the classification ability of a covering approximation space to reduce the relatively reducible elements in it. In addition, combining absolute reduction and relative reduction, an algorithm for knowledge reduction of covering approximation space is developed. It can reduce not only absolutely redundant knowledge, but also relatively redundant knowledge. It is significant for the following-up steps of data mining.

Yingxu Wang investigates "Formal Description of the Cognitive Process of Memorization." Memorization is a key cognitive process of the brain because almost all human intelligence functioning is based on it. This paper presents a neuroinformatics theory of memory and a cognitive process of memorization. Cognitive informatics foundations and functional models of memory and memorization are explored toward a rigorous explanation of memorization. The cognitive process of memorization is studied, revealing how and when memory is created in long-term memory. On the basis of the formal memory and memorization models, the cognitive process of memorization is rigorously described using real-time process algebra (RTPA). This work is one of the fundamental enquiries into the mechanisms of the brain and natural intelligence according to the layered reference model of the brain (LRMB) developed in cognitive informatics.

Andrew Gleibman presents "Intelligent Processing of an Unrestricted Text in First Order String Calculus." First order string calculus (FOSC), introduced in this paper, is a generalization of first order predicate calculus (FOPC). The generalization step consists in treating the unrestricted strings, which may contain variable symbols and a nesting structure, similarly to the predicate symbols in FOPC. As a logic programming technology, FOSC, combined with a string unification algorithm and the resolution principle, eliminates the need to invent logical atoms. An important aspect of the technology is the possibility to apply a matching of the text patterns immediately in logical reasoning. In this way the semantics of a text can be defined by string examples, which only demonstrate the concepts, rather than by a previously formalized mathematical knowledge. The advantages of avoiding this previous formalization are demonstrated. The author investigates the knowledge representation aspects, the algorithmic properties, the brain simulation aspects, and the application aspects of FOSC theories in comparison with those of FOPC theories. FOSC is applied as a formal basis of logic programming language Sampletalk, introduced in earlier publications.

Xia Wang and Wenxiu Zhang study "Knowledge Reduction in Concept Lattices Based on Irreducible Elements." As one of the important problems of knowledge discovery and data analysis, knowledge reduction can make the discovery of implicit knowledge in data easier and the representation simpler. In this paper, a new approach to knowledge reduction in concept lattices is developed based on irreducible elements, and characteristics of attributes and objects are also analyzed. Furthermore, algorithms for finding attribute and object reducts are provided. The algorithm analysis shows that the approach to knowledge reduction involves less computation and is more tractable compared with the current methods.

Yousheng Tian, Yingxu Wang, and Kai Hu present "A Knowledge Representation Tool for Autonomous Machine Learning Based on Concept Algebra." Concept algebra is an abstract mathematical structure for the formal treatment of concepts and their algebraic relations, operations, and associative rules for composing complex concepts, which provides a denotational mathematic means for knowledge system representation and manipulation. This paper presents an implementation of concept algebra by a set of simulations in Java. A visualized knowledge representation tool for concept algebra is developed, which enables machines to learn concepts and knowledge autonomously. A set of eight relational operations and nine compositional operations of concept algebra are implemented in the tool to rigorously manipulate knowledge by concept networks. The knowledge representation tool is capable of presenting concepts and knowledge systems in multiple ways in order to simulate and visualize the dynamic concept networks during machine learning based on concept algebra.

Robert Harrison and Christine W. Chan develop "Dyna: A Tool for Dynamic Knowledge Modeling." This paper presents the design and implementation of an ontology construction support tool. The inferential modeling technique (IMT), which is a technique for modeling the static and dynamic knowledge elements of a problem domain, provided the basis for the design of the tool. Existing tools lack support for modeling dynamic knowledge as defined by the IMT. Therefore, the focus of this work is the development of a Protégé plug-in, called Dyna, which supports dynamic knowledge modeling and testing. Within Dyna, the task behavior language (TBL) supports formalized representation of the task behavior component of dynamic knowledge. The interpreter for TBL can also enable the task behavior representation to be run and tested, thus enabling verification and testing of the model. Dyna also supports storing the dynamic knowledge models in XML and OWL so that they can be shared and re-used across systems. The tool is applied for constructing an ontology model in the domain of petroleum contamination remediation selection.

Dominik Slezak presents "Rough Sets and Functional Dependencies in Data: Foundations of Association Reducts." The author investigates the notion of association reducts that represent data-based functional dependencies between sets of attributes. It is preferred in the reducts that possibly smaller sets determine larger ones. An association reduct is compared with other types of reducts previously studied within the theory of rough sets. The author focuses particularly on modeling inexactness of dependencies, which is crucial for many real-world applications. In addition, the optimization problems and algorithms are studied that aim to search for the most interesting approximate association reducts in data.

Anjali Mahajan and M.S. Ali study a "Hybrid Evolutionary Algorithm for the Graph Coloring Register Allocation Problem for Embedded Systems." Memory or registers are used to store the results of computation of a program. As compared to memory, accessing a register is much faster, but they are scarce resources in real-time embedded systems and have to be utilized very efficiently. If the register set is not sufficient to hold all program variables, certain values have to be stored in memory and the so-called spill code has to be inserted. The optimization goal is to hold as many live variables as possible in registers in order to avoid expensive memory accesses. The register allocation phase is generally more challenging in embedded systems. In this paper, the authors present a new hybrid evolutionary algorithm (HEA) for graph coloring register allocation problems for embedded systems based on a new crossover operator called crossover by conflict-free sets (CCS) and a new local search function. The objective is to minimize the total spill cost.

Huawen Liu, Jigui Sun, Huijie Zhang, and Lei Liu present "Extended Pawlak's Flow Graphs and Information Theory." Flow graph is an effective graphical tool of knowledge representation and analysis. It explores dependent relations between knowledge in the form of information flow quantity. However, the quantity of flow cannot exactly represent the functional dependency between knowledge. In this paper, the authors present an extended flow graph using concrete information flow, and then give its interpretation under the framework of information theory. Subsequently, an extended flow graph generation algorithm based on the significance of attribute is proposed in virtue of mutual information. In addition, for the purpose of avoiding over-fitting and reducing store space, a reduction method about this extension using information metric is also developed.

The editors expect that the readers of the journal of *Transactions on Computational Science* (TCS), Vol. V, will benefit from the papers presented in this special issue on the latest advances in cognitive knowledge representation, processing, and their applications in cognitive informatics, cognitive computing, computational intelligence, and AI.

The guest editors of this *Special Issue on Cognitive Knowledge Representation* in the Springer *Transactions on Computational Science* V would like to thank all authors for submitting their interesting work. We are grateful to the reviewers for their great contributions to this special issue. We would like to express our sincere appreciation to the Editors-in-Chief, Marina L. Gavrilova and Chih Jeng Kenneth Tan, for their advice and support. We also thank the editorial staff at Springer for their professional help during the publication of this special issue.

February 2009   Yingxu Wang
                                                                Keith C.C. Chan

LNCS Transactions on Computational Science – Editorial Board

Table of Contents

Toward a Formal Knowledge System Theory and Its Cognitive Informatics Foundations

Yingxu Wang

Theoretical and Empirical Software Engineering Research Centre (TESERC)
International Center for Cognitive Informatics (ICfCI)
Dept. of Electrical and Computer Engineering
Schulich School of Engineering, University of Calgary
2500 University Drive, NW, Calgary, Alberta, Canada, T2N 1N4
Tel.: (403) 220 6141; Fax: (403) 282 6855
yingxu@ucalgary.ca

Abstract. Knowledge science and engineering are an emerging field that studies the nature of human knowledge, and its manipulations such as acquisition, representation, creation, composition, memorization, retrieval, and depository. This paper presents the nature of human knowledge and its mathematical models, internal representations, and formal manipulations. The taxonomy of knowledge and the hierarchical abstraction model of knowledge are investigated. Based on a set of mathematical models of knowledge and the Object-Attribute-Relation (OAR) model for internal knowledge representation, rigorous knowledge manipulations are formally described by concept algebra. A coherent framework of formalized knowledge systems is modeled based on the analyses of roles of formal and empirical knowledge. Then, the theory of knowledge acquisition and the cognitive model of knowledge spaces are systematically developed.

Keywords: Cognitive informatics, cognitive computing, knowledge science, knowledge engineering, properties of knowledge, generic information model, knowledge creation, acquisition, manipulation, modeling, abstraction levels, formal knowledge system, LRMB, HAM, OAR, computational intelligence.

1 Introduction

One of the most famous assertions of Francis Bacon (1561-1626) is that *"knowledge is power."* Knowledge is acquired information in forms of data, behavior, experience, and skills retained in memory through learning. Knowledge science and engineering are an emerging field that studies the nature of human knowledge, its mathematical models, formal representations, and manipulations. Because almost all disciplines of sciences and engineering deal with information and knowledge, investigation into the generic theories of knowledge science and its cognitive foundations is one of the profound areas of cognitive informatics.

Any knowledge acquired has to be represented and retained in memory of the brain. The human memory encompasses the Sensory Buffer Memory (SBM), Short-Term Memory (STM), Long-Term Memory (LTM) [1], [7], [8], [10], [11], [13], [14], as well

M.L. Gavrilova et al. (Eds.): Trans. on Comput. Sci. V, LNCS 5540, pp. 1–19, 2009.

as Action Buffer Memory (ABM) and Conscious-Status Memory (CSM) [32], [33]. Among these memories, LTM is the permanent memory that human beings rely on for storing acquired information such as facts, knowledge and experiences. However, the main part of skills and behaviors are stored in ABM as logically modeled in [32], which is interconnected with the motor servo muscles.

Corresponding to the forms of memories in the brain, human knowledge as cognized or comprehended information can be defined in the narrow and broad senses.

Definition 1. *Knowledge*, in the narrow sense, is acquired information in LTM or acquired skills in ABM through learning.

In contrary, the broad sense of knowledge refers to the creative products of natural intelligence generated in any form of acquired memory such as abstract knowledge, behaviors, experience, and skills.

Definition 2. *Knowledge*, in the broad sense, is acquired information in forms of abstract knowledge, behavior, experience, and skills through learning in LTM or ABM.

According to Definitions 1 and 2, the concept of knowledge has a close connection with the concepts of memory [20], [28] and learning [21]. *Memory* is the physiological organs or networked neural clusters in the brain for retaining and retrieving information. Therefore, the contents of memory are all forms of knowledge, such as data, behavior, experience, and skills. *Memorization* is a cognitive process of the brain at the meta-cognitive layer that establishes (encodes and retains) and reconstructs (retrieves and decodes) information in LTM. *Learning* is a cognitive process of the brain at the higher cognitive layer that gains knowledge about something or acquires skills for some actions by updating the entire cognitive models of the brain in LTM.

Definition 3. Knowledge science is an emerging field that studies the nature of human knowledge, and its manipulations such as acquisition, representation, creation, composition, memorization, retrieval, and depository.

Definition 4. Knowledge engineering is applied knowledge science in the design and implementation of knowledge systems and the organization and manipulation of knowledge applications.

This paper explores the theoretical foundations of knowledge science and engineering. The neural informatics foundations of knowledge are presented in Section 2, which covers the Object-Attribute-Relation (OAR) model for knowledge representation, taxonomy of knowledge, and the layered abstraction model of knowledge. The mathematical model of knowledge is created in Section 3 with the abstract concept model and concept algebra. The framework of formal knowledge systems and the comparative analyses of formal and empirical knowledge are presented in Sections 4. Then, theories of knowledge acquisition are described in Section 5 with the effort and complexity models of knowledge creation and acquisition.

2   The Hierarchical Abstraction Model of Knowledge

This section describes the taxonomy of knowledge and its abstract levels by a hierarchical abstraction model of knowledge. More rigorous mathematical models of knowledge representation and manipulation will be developed in Section 3.

2.1 Taxonomy of Knowledge

All science and engineering disciplines deal with knowledge. However, in literature, data, information, and knowledge were considered different [2], [34]. It is usually perceived that *data* are directly acquired raw information, which is a quantitative abstraction of external objects and/or their relations. *Information* is meaningful data or the subjective understanding of data. At the top level, *knowledge* is consumed information and data related to those have already existed and accumulated in the brain.

Based on the investigations in cognitive informatics [15], [16], [17], [19], [22], [26], [29], particularly the study on the OAR model and the mechanisms of internal information representation [20], the above empirical classification of the cognitive levels of data, information, and knowledge may be revised. With a unified perspective, data (sensational inputs), actions (behavioral outputs), and their internal representations in forms of knowledge, experience, and skill are all cognitive information. Types of cognitive information may be classified on the basis of how the internal information related to the input and output of the brain as shown in Table 1.

Table 1. Taxonomy of Knowledge

		Type of output		Ways of acquisition
		Information	Action	
Type of input	**Information**	Knowledge (*K*)	Behavior (*B*)	*Direct or indirect*
	Action	Experience (*E*)	Skill (*S*)	*Direct only*

As shown in Table 1, the taxonomy of cognitive information is determined by the types of inputs and outputs of information to and from the brain, where both inputs and outputs can be either information (I) or action (A).

Definition 5. The *Generic Information Model* (GIM) classifies cognitive information into four categories, according to their types of I/O information, known as *knowledge, behavior, experience,* and *skill*, i.e.:

$$\text{a) Knowledge} \quad K: I \rightarrow I \tag{1}$$
$$\text{b) Behavior} \quad B: I \rightarrow A \tag{2}$$
$$\text{c) Experience} \quad E: A \rightarrow I \tag{3}$$
$$\text{d) Skill} \quad S: A \rightarrow A \tag{4}$$

According to the GIM model, for a given cognitive process, if both I/O are abstract information, the internal information acquired is *knowledge*; if both I/O are empirical actions, the type of internal information is *skill*; and the remainder combinations between action/information and information/action produce *experience* and *behaviors*, respectively. It is noteworthy in Definition 5 and Table 1 that behaviors is a new type of cognitive information modeled inside the brain, which embodies an abstract input to an observable behavioral output [19].

Although knowledge or behaviors may be acquired directly or indirectly, skills and experiences can only be obtained directly by hand on activities. Further, the storage locations of the abstract information are different, where knowledge and experience are stored as abstract relations in LTM, while behaviors and skills are stored as wired neural connections in ABM.

Theorem 1. *Knowledge* and *behaviors* can be learnt indirectly by inputting abstract information; while *experience* and *skills* must be learnt directly by hands-on or empirical actions.

Storage forms of knowledge in memories can be summarized in Table 2 [6], [18], [20], [34], where the interrelationships between the forms of memories and the forms of knowledge are created. For example, knowledge is retained in the form of semantic memory, experience in episodic memory, skill in procedural memory, and behavior in procedural/action buffer memory. Rigorous models may be referred to neuroinformatics [19].

The classification of internal information in cognitive informatics can be used to explain a wide range of phenomena of learning and practices. For instance, it explains why people have to make the same mistakes in order to gain empirical experiences and skills, and why experience transfer would be so hard and could not be gained by indirect reading.

Table 2. Functional Categorization of Knowledge

Category	Subcategory	Type in GIM	Description
Abstract knowledge (To be)	Semantic	$K: I \rightarrow I$	General objective facts and knowledge about the physical and perceived worlds
	Episodic	$E: A \rightarrow I$	Experienced events or sequence of events
Behavioral knowledge (To do)	Imperative	$B: I \rightarrow A$	Conditioned reactions or instructive actions
	Procedural	$S: A \rightarrow A$	Acquired skills

2.2 The Abstract Levels of Knowledge

Abstraction is a gifted capability of mankind. Abstraction is a basic cognitive process of the brain at the meta-cognitive layer according to the Layered Reference Model of the Brain (LRMB) [32]. Only by abstraction important principles and properties about given objects under study may be elicited and discovered from a great variety of phenomena and empirical observations in an area of inquiry.

Definition 6. *Abstraction* is a process to elicit a subset of objects in a given discourse that shares a common property, which may be used to identify and distinguish the subset from the whole in order to facilitate reasoning.

Abstraction is a powerful means of philosophy and mathematics. All formal logical inferences and reasonings may only be carried out on the basis of common and abstract properties shared by a given set of objects under study.

Theorem 2. The *principle of abstraction* states that, given any set S and any property p, abstraction is to elicit a subset E such that the elements of it, e, possess the property $p(e)$, i.e.:

$$\forall S, p \Rightarrow \exists E \subseteq S \wedge \forall e \in E, p(e) \tag{5}$$

According to the Information-Matter-Energy (IME) model [19], there are two categories of objects under studies in science and engineering known as the *concrete* entities in the real world and the *abstract* objects in the information world. In the latter, an

Table 3. Abstract Levels of Knowledge and Cognitive Information

Level	Category	Description
L1	Analogue objects	Real-world entities, empirical artifacts
L2	Diagrams	Geometric shapes, abstract entities, relations
L3	Natural languages	Empirical methods, heuristic rules
L4	Professional notations	Special notations, rigorous languages, formal methods
L5	Mathematics (philosophy)	High-level abstraction of objects, attributes, and their relations and rules, particularly those that are time and space independent

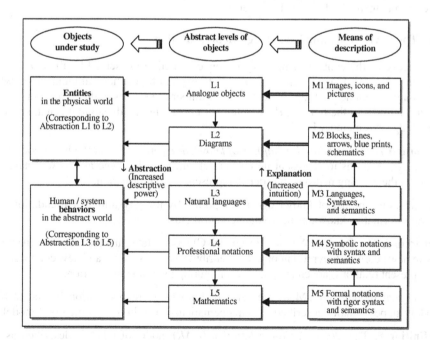

Fig. 1. The Hierarchical Abstraction Model (HAM) of knowledge and information

important part of the abstract objects is human or system behaviors, which are planned or executed actions onto the real-world entities and abstract objects. The abstract levels of cognitive information of both the objects and their behaviors can be divided into five levels as shown in Table 3.

Theorem 3. The *Hierarchical Abstraction Model* (HAM) of knowledge states that the extend of abstraction of cognitive information can be classified at five levels such as those of *analogue objects, diagrams, natural languages, professional notations,* and *mathematics.*

The HAM model is illustrated in Fig. 1 with five abstract levels where each level is corresponding to a certain descriptive means. The higher the abstraction level of an object, the more complicated the description means.

There are two approaches to system description and modeling as shown in the HAM model known as *abstraction* and *explanation.* The former enables the enhancement of the descriptive power in terms of expressiveness, precise, and rigor; while the latter enables the improvement of the intuitiveness of understanding and comprehension using a means much closer to the real world images and analogue objects directly acquired by the sensations of the brain.

3 Mathematical Model of Knowledge

On the basis of the HAM model of knowledge developed in preceding section, this section formalizes the abstract model of knowledge and their internal representation by concept algebra and the OAR model.

3.1 The OAR Model for Knowledge Representation

Knowledge as the content of memory is conventionally represented by the *container* metaphor [1], [4]. [34], which was not able to explain how a huge amount of knowledge may be retained without increasing the number of neurons in the brain. This subsection introduces the *relational* metaphor for explaining the dynamic mechanisms of knowledge and memory [20].

Definition 7. The *relational metaphor of knowledge* perceives that the brain does not create new neurons to represent new information, instead it generates new synapses between the existing neurons in order to represent new information.

The relational model of knowledge is supported by the physiological foundation of neural informatics known as the dynamic neural cluster model.

Theorem 4. The *Dynamic Neural Cluster* (DNC) model states that the LTM is dynamic, where new neurons (to represent objects or attributes) are assigning, and new connections (to represent relations) are creating and reconfiguring over time in the brain.

The logical model of LTM or the form of knowledge representation by the DNC model can be formally described by a mathematical model known as the OAR model.

Definition 8. The *Object-Attribute-Relation* (OAR) model of LTM is described as a triple, i.e.:

$$OAR \triangleq (O, A, R) \tag{6}$$

where O is an abstraction of an external entity and/or internal concept, A is a sub-object that is used to denote detailed properties and characteristics of the given object, and R is a connection or inter-relationship between any pair of object-object, object-attribute, and attribute-attribute.

According to the OAR model, the result of knowledge acquisition or learning can be embodied by the updating of the existing OAR in the brain. In other words, learning is a dynamic composition of the existing OAR in LTM and the currently created sub-OAR as expressed below.

Theorem 5. The *entire knowledge model* maintained in the brain, or the representation of learning results, states that the internal memory in the form of the OAR structure can be updated by a composition ⊎ between the existing OAR and the newly created sub-OAR (*sOAR*), i.e.:

$$OAR\text{'ST} \triangleq OAR\text{ST} \uplus sOAR\text{'ST}$$

$$= OAR\text{ST} \uplus (O_s, A_s, R_s) \tag{7}$$

The above theory of knowledge acquisition lays an important foundation for learning theories and pedagogy.

3.2 Modeling and Manipulating Knowledge by Concept Algebra

A *concept* is a basic cognitive unit of knowledge and inferences [5], [9], [23], [24] by which the meanings and semantics of a real-world entity or an abstract entity may be represented and embodied based on the OAR model [20]. According to concept algebra [24], the mathematical model of a generic abstract concept is as follows.

Definition 9. An *abstract concept c* is a 5-tuple, i.e.:

$$c \triangleq (O, A, R^c, R^i, R^o) \tag{8}$$

where

- O is a finite nonempty set of objects of the concept, $O = \{o_1, o_2, ..., o_m\} \subseteq \mathbb{P}\mathcal{O}$, where $\mathbb{P}\mathcal{O}$ denotes a power set of $\mathcal{O}$.
- A is a finite nonempty set of attributes, $A = \{a_1, a_2, ..., a_n\} \subseteq \mathbb{P}\mathcal{A}$.
- $R^c = O \times A$ is a set of internal relations.
- $R^i \subseteq A' \times A$, $A' \sqsubseteq C' \wedge A \sqsubseteq c$, is a set of input relations, where C' is a set of external concepts, $C' \subseteq \Theta_C$. For convenience, $R^i = A' \times A$ may be simply denoted as $R^i = C' \times c$.
- $R^o \subseteq c \times C'$ is a set of output relations.

Any real-world and concrete concept can be rigorously modeled using the abstract concept model. Further, a set of algebraic operations can be defined on abstract concepts, which form a mathematical structure known as concept algebra.

Definition 10. A *concept algebra CA* on a given semantic environment Θ_C is a triple, i.e.:

$$CA \triangleq (C, OP, \Theta_C) = (\{O, A, R^c, R^i, R^o\}, \{\bullet_r, \bullet_c\}, \Theta_C) \tag{9}$$

where $OP = \{\bullet_r, \bullet_c\}$ are the sets of *relational* and *compositional* operations on abstract concepts.

Definition 11. The *relational operations* $\bullet_r$ in concept algebra encompass 8 comparative operators for manipulating the algebraic relations between concepts, i.e.:

$$\bullet_r \triangleq \mathcal{R} = \{\leftrightarrow, \nleftrightarrow, \prec, \succ, =, \cong, \sim, \triangleq\} \tag{10}$$

where the relational operators stand for *related, independent, subconcept, superconcept, equivalent, consistent, comparison,* and *definition*, respectively.

Definition 12. The *compositional operations* $\bullet_c$ in concept algebra encompass 9 associative operators for manipulating the algebraic compositions among concepts, i.e.:

$$\bullet_c \triangleq \Gamma = \{\Rightarrow, \xrightarrow{-}, \xrightarrow{+}, \xrightarrow{\sim}, \uplus, \sqcap, \Leftarrow, \vdash, \mapsto\} \tag{11}$$

where the compositional operators stand for *inheritance, tailoring, extension, substitute, composition, decomposition, aggregation, specification,* and *instantiation*, respectively.

Concept algebra deals with the algebraic relations and associational rules of abstract concepts. The associations of concepts form a foundation to denote complicated relations between concepts in knowledge representation. Detailed definitions of the algebraic operations in concept algebra have been provided in [24].

On the basis of concept algebra, a generic model of human knowledge may be rigorously described as follows.

Definition 13. A *generic knowledge K* is an *n*-nary relation R_k among a set of *n* concepts in *C*, i.e.:

$$K = R_k : (\underset{i=1}{\overset{n}{X}} C_i) \rightarrow C \tag{12}$$

where $\underset{i=1}{\overset{n}{U}} C_i = C$, and $R_k \in \Gamma$.

In Definition 13 the relation R_k is one of the concept operations defined in concept algebra [24] that serves as the knowledge composing rules.

A complex knowledge is a composition of multiple concepts in the form of a concept network.

Definition 14. A *concept network CN* is a hierarchical network of concepts interlinked by the set of nine associations $\mathfrak{R}$ defined in concept algebra, i.e.:

$$CN = R_k : \underset{i=1}{\overset{n}{X}} C_i \rightarrow \underset{i=j}{\overset{n}{X}} C_j \tag{13}$$

where $R_k \in \Gamma$.

Because the relations between concepts are transitive, the generic topology of knowledge is a hierarchical concept network. The advantages of the hierarchical knowledge architecture in the form of concept networks are as follows:

    a) *Dynamic*: The knowledge networks may be updated dynamically along with information acquisition and learning without destructing the existing concept nodes and relational links.

    b) *Evolvable*: The knowledge networks may grow adaptively without changing the overall and existing structure of the hierarchical network.

4   The Framework of Formalized Knowledge Systems

Knowledge and information are the essence of nature and the foundation for mankind evolution. This section contrasts formal and empirical knowledge, and analyzes the roles of knowledge in mankind evolution. Then, a formal knowledge framework will be established.

4.1   Roles of Knowledge in Mankind Evolution

It is recognized that the basic evolutional need of mankind is to preserve both the species' biological traits and the cumulated information/knowledge bases. For the former, gene pools are adopted to pass human trait information via DNA from generation to generation. However, for the latter, because acquired knowledge cannot be physiologically inherited between generations and individuals, various information means and systems are adopted to pass cumulated human information and knowledge.

It is noteworthy that intelligence plays an irreplaceable role in the transformation between information, matter, and energy [19]. It is observed that almost all cells in human bodies have a certain lifecycle in which they reproduce themselves via divisions. This mechanism allows human trait information to be transferred to offspring through gene (DNA) replications during cell reproduction. However, it is observed that the most special mechanism of neurons in the brain is that they are the only type of cells in human body that does not go through reproduction but remains alive throughout the entire human life [4], [14]. The advantage of this mechanism is that it enables the physiological representation and retention of acquired information and knowledge to be memorized in long-term memory. But the key disadvantage of this mechanism is that it does not allow acquired information to be physiologically passed on to the next generation, because there is no DNA replication among memory neurons.

This physiological mechanism of neurons in the brain explains not only the foundation of memory and memorization, but also the wonder why acquired information and knowledge cannot be passed and inherited physiologically through generation to generation. Therefore, to a certain extent, mankind relies very much on information for evolution than that of genes, because the basic characteristic of the human brain is intelligent information processing. In other words, the intelligent ability to cumulate and transfer knowledge from generation to generation plays the vital role in mankind's evolution for both individuals and the entire species. This distinguishes human beings from other species in natural evolution, where the latter cannot systematically pass acquired knowledge by external and persistent information systems from generation to generation in order to enable it to grow cumulatively and exponentially.

4.2 Roles of Formal and Empirical Knowledge

It is recognized that the entire knowledge of mankind may be classified into *empirical* and *theoretical* categories. The former is the direct knowledge about the physical world; while the latter is the derived knowledge about both the physical and abstract worlds. In contrasting the nature of empirical knowledge and theoretical knowledge, the following principle on the basic properties of knowledge can be derived.

Theorem 6. The *rigorous levels of empirical and theoretical knowledge* states that an *empirical truth* is a truth based on or verifiable by observations, experiments, or experiences. In contrary, a *theoretical proposition* is an assertion based on formal theories or logical inferences.

Empirical knowledge answers *how*; while theoretical knowledge reveals *why*. Theoretical knowledge is a formalization of generic truth and proven empirical knowledge. Although the discovery and development of a theory or a law may take decades even centuries, its acquisition and exchange are much easier and faster by ordinary effort. However, empirical knowledge is very difficult to be gained. One may consider that a person can learn multiple scientific disciplines, but few persons may become an expert in multiple engineering disciplines such as in all areas of electrical, mechanical, chemical, and computer engineering. The reasons behind this are that each engineering area requires specific empirical knowledge, skills, and tools. All of them need a long period of training and practice to acquire.

Huge empirical knowledge were created and disappeared over time. For example, there are a lot of empirical knowledge on software engineering published each year in the last decades. However, those would be included in a textbook on software engineering theories or foundations as proven and general truth, rather than specific cases partially working on certain given or non-specified conditions, would be listed in only a few pages.

A major risk of empirical knowledge is its uncertainty when it is applied in a different environment, even the same environment but at different time, because empirical knowledge and common senses are often error-prone. For instance:

- In early age of human civilization, people were commonly believed that our earth is flat, and we lived in the center of the universe, until Nicholas Copernicus (1473-1543) proven these common senses were false in the early 16th century.
- Managers believed that the larger the project, the larger the team required. However, Wang's *coordinative work organization theory* [18], formally reveals that for a given workload, the optimal labor allocation and the shortest project period are constrained by the interpersonal coordination rate r. That is, putting more than the optimal number of persons into the project is not only counterproductive, but also dramatically increasing the real workload of the project – a major hidden reason of most project failures in coordination-intensive projects such as software engineering.

Therefore, empirical knowledge needs to be refined and formalized into theoretical knowledge. It is a common observation that all recurrent objects in nature and their relations are constrained by certain invariant laws, no matter one observed them or not at a given time. This is one of the essences that may be gained in knowledge science and engineering.

4.3 A Coherent Framework of Formal Knowledge Systems

A formal knowledge system is needed to maintain a stable, efficient, and rigorous inference base, in which logic reasoning may be conducted. Mathematical science provides a successful paradigm to organize and validate human knowledge, where once a truth or a theorem is established, it is true till the axioms or conditions that it is based are changed or extended. Therefore, a theorem does not need to be argued each time when one applies in reasoning. This is the advantage and efficiency of formal knowledge in sciences and engineering. In other words, if any theory or assertion may be argued from time-to-time based on a wiser idea or a trade-off, it is an empirical result rather than a formal theory in knowledge.

The framework of a Formal Knowledge System (FKS) can be described as shown in Fig. 2. The FKS framework demonstrates the interrelationships between a comprehensive set of terms of formal knowledge, where the taxonomy of formal knowledge and their definitions are presented in Table 4.

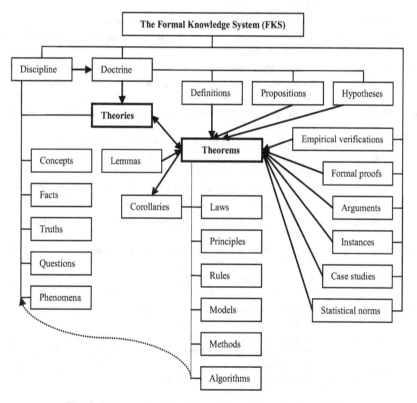

Fig. 2. Framework of the Formal Knowledge System (FKS)

Table 4. Taxonomy of Formal Knowledge Systems

No	Term	Description
1	Algorithm	A generic and reusable method described by rules or processes in problem-solving.
2	Argument	A reason or a chain of reasons on the truth of a proposition or theory.
3	Case study	An applied study of generic theory in a particular setting or environment.
4	Concept	A cognitive unit in reasoning by which the meaning and semantics of real-world or abstract entities may be represented and embodied.
5	Corollary	A proposition that follows from or appended to a theorem has already been proven.
6	Definition	An exact or formal description of a concept or fact as a basis of reasoning.
7	Discipline	A branch of knowledge that studies a category of objects with a set of doctrines, frameworks, theories, and methodologies.
8	Doctrine	A set of coherent theories.
9	Empirical verification	A proof of a truth, accuracy, or validity of a proposition or theory based on observation and experience.
10	Factor	A thing or relation that is observed or proven true.
11	Hypothesis	A proposed proposition as a basis for reasoning or investigation in order to prove its truth or falsity.
12	Instance	An example or particular case of a general phenomenon.
13	Law	A proven statement of a causality between a deducted phenomenon or variable and its conditions.
14	Lemma	A subsidiary or intermediate theorem in a chain of arguments or proofs.
15	Method	An established procedure or approach to solve a class of problems, or to carry out a kind of tasks.
16	Model	A description of an architecture, mechanism, and/or behavior of a system or process.
17	Phenomenon	An observed fact or state with known or unknown causality.
18	Principle	A generalized axiom or proposition that explains a wide range of cases or instances in a field of study.
19	Proof	An established fact or validated statement by evidences and arguments.
20	Proposition	A formal statement of an assertion of judgment or a problem.
21	Question	A doubt about the truth of a proposition, or a request for a solution to a problem.
22	Rule	A proposition that describes or prescribes allowable conditions and domains of a law or principle.
23	Statistical norm	A typical, average, or standard quality, quantity, or state of a phenomenon or system based on a large set of observations and statistic analyses.
24	Theorem	A generic proposition expressed formally and established by means of accepted truths.
25	Theory	A system of generic and formalized principles, theorems, laws, relations, and models independent of objects to be explained or practices are based.
26	Truth	An established, proven, or accepted constant state of a term or proposition in reasoning that is either true or false.

The FKS model is centered by a set of theories, where a *theory* is a statement of how and why certain objects, facts, or truths are related. Theoretical knowledge is a formalization of generic truth and proven empirical knowledge. According to Lemmas 1 and 2, theoretical knowledge may be easier to acquire when exist. However, empirical knowledge is very difficult to be gained without hand-on practice.

According to the FKS framework, an immature discipline of science and engineering is characterized by that its body of knowledge is not formalized or is mainly empirical. When there is no theory in a field of human enquiry, the practice in it is risk-prone. Instead of a set of proven theories, immature disciplines document a large set of observed facts, phenomena, and their possible or partially working explanations. In such disciplines, researchers and practitioners might be able to argue every informal assertion documented in natural languages from time-to-time, until it is formalized in mathematical forms and proven rigorously.

The disciplines of mathematics and physics are successful branches of sciences that adopt the formal knowledge system. The advantages of FKS's are their *stability* and *efficiency*. The former is a property of formal knowledge that once it is established and proven, users who refers to it will no longer need to reexamine or reprove it. The latter is a property of formal knowledge that is exclusively true or false in reasoning that forms a set of solid foundations in reasoning.

5   The Theory of Knowledge Acquisitions

Knowledge creation and acquisition can be elaborated on the basis of studies in cognitive informatics [15], [19] and denotational mathematics [23]. An analysis of the cognitive model of knowledge spaces will reveal the impact of multidisciplinary knowledge in knowledge engineering.

5.1   The Effort Model of Knowledge Creation and Acquisition

According to the OAR model and Definitions 13 and 14, a generic knowledge is a relation between two or more abstract concepts in LTM, while a skill is a relation between a concept and an action or behavior in ABM. In other words, knowledge represents what *to be*, while skills denote what *to do*.

It is observed that human knowledge creation and development is so difficult where solving a hard problem often costs an effort of years, decades, even centuries. However, once the problem is solved and the knowledge is created, only an ordinary effort is needed by individuals to understand and acquire it fairly quickly with no difficulty. This phenomenon in knowledge science can be described more formally as follows.

Lemma 1. For a specific new knowledge K, the *effort spent in its creation* $E_c(K)$ is far more greater than that of its acquisition $E_a(K)$, i.e.:

$$E_c(K) >> E_a(K) \tag{14}$$

It is noteworthy that the creation of knowledge is a conservative process that establishes a novel relation between two or more objects or concepts by searching and evaluating a vast space of possibilities in order to explain a set of natural phenomena or abstract problems [27]. Since the memory capacity of human can be as high as $10^{8,432}$ bits as estimated in [31], the complexity in search for new knowledge is necessarily infinitive, if not a short cut should be discovered by chance during extensive and persistent thoughts. However, the acquisition of knowledge is to simply add a known relation in the LTM of an existing knowledge structure. This is why the effort for acquiring an existing knowledge is very low than that of knowledge creation.

On the other hand, the speed for acquiring skills and experiences may be much slower than that of knowledge acquisition, because of the need of hands on actions and the creation of a permanent internal performance model in ABM that requires highly repetitive practice.

Lemma 2. The *effort of skill acquisition $E_a(S)$* is far more greater than that of knowledge acquisition $E_a(K)$, i.e.:

$$E_a(S) \gg E_a(K) \tag{15}$$

Another angle to analyze the effort of knowledge creation in science and engineering can be by estimating the workload as developed in the coordinative work organization theory [18] as given in the following theorem.

Theorem 7. The *law of coordinative work organization* states that the *actual workload W* of a coordinative project is a function of the average interpersonal coordination rate r and the number of labor L in the project, i.e.:

$$
\begin{aligned}
W &= L \bullet T \\
&= L \bullet T_1(1 + h) \\
&= W_1(1 + h) \\
&= W_1(1 + r \bullet \frac{L(L-1)}{2}) \quad \text{[PM]}
\end{aligned}
\tag{16}
$$

where T_1 is the *indicational duration* needed to complete the work by one person only, and W_1 is the *ideal workload* without the interpersonal overhead h or that of a single person project.

Example 1. During authoring the textbook on "*Software Engineering Foundations: A Software Science Perspective* [18]", knowledge across 12 related science and engineering disciplines, such as philosophy, mathematics, computing, cognitive informatics, system science, management science, and economy, have been investigated. The statistics data on the workload of the book can be estimated below with the time spent known as $T \approx 10,000$ hrs = 42.0 months, i.e.:

$$
\begin{aligned}
W &= W_1 = L \bullet T \\
&= 1 \bullet 42.0 \\
&= 42.0 \text{ [PM]}
\end{aligned}
$$

If this book were authored by multiple authors from 12 different disciplines aiming at the same degree of seamless coherency and consistency as that of this solely authored book, rather than an edited assembly of individual views, the effort according to Theorem 7 would be the following subject to $r = 30\%$:

$$W = W_1(1 + r \bullet \frac{L(L-1)}{2})$$

$$= 42.0 \ (1 + 0.3 \bullet \frac{12.0(12.0 - 1)}{2})$$

$$= 873.6 \ [PM]$$

The result is 837.6 person-month or about 72.8 person-years for the given coordination overhead $r = 30\%$, which indicates a mission virtually impossible. This example shows that highly complicated problems may be feasibly resolved by a single brain with enhanced and necessary multidisciplinary knowledge, rather than by a group of individual counterparts. That is why a software engineering project should not involve too many architects in the early phase of system design and modeling, no matter how complicated it is.

5.2 The Complexity Model of Knowledge Creation

According to the relational complexity theory of systems [25], [30], the domain and magnificent of experts' knowledge can be estimated based on the number of abstract objects or concepts and relations among them.

Definition 15. The potential *number of relations* among the combination of knowledge from n disciplines, $C_r(n)$, is in the order of n^2, i.e.:

$$C_r(n) = O(n^2)$$
$$= n \bullet (n\text{-}1) \qquad (17)$$

where it is assumed that a pairwise relation r is asymmetric, that is, $r(a, b) \neq r(b, a)$.

Example 2. For the book as described in Example 1 that covered fundamental theories of 12 disciplines, assuming each discipline has 1,000 concepts in average, the total number of consumed concepts that an expert needs to acquire a cohesive knowledge with all the 12 disciplines, $C_r(n)$, is up to the following:

$$C_r(n) = n \bullet (n\text{-}1)$$
$$\approx 12 \bullet (12\text{-}1) \bullet 10^6$$
$$= 1.32 \times 10^8 \qquad (18)$$

This figure shows that totally about 132 million new relations between multidisciplinary concepts need to be generated in the brain before and during the written of the book on multidisciplinary foundations. In other words, the relational complexity of a multidisciplinary knowledge system (Eq. 17) is huge enough to enable new concepts, principles, and theories to be created, which may not belongs to any individual disciplines but at the edge of them. This reveals the advantages of an expert who possesses multidisciplinary knowledge toward a set of intricate problems under study.

Constrained by the cognitive, organizational, and resources limitations and their complicated interrelations, most persistent problems in modern sciences and engineering are not a trivial one. If only empirical studies are conducted, perhaps many additional decades are still needed to accidentally find a solution for such problems.

5.3 The Cognitive Model of Knowledge Spaces

As explained in Section 5.2, the most interesting problems in research are at the edges of conventional disciplines. Therefore, transdisciplinary and multidisciplinary research are necessary. In addition, the maintenance of a global and holistic view is one of the fundamental insights of scientific research.

Lemma 3. The impact of an expert with coherently m disciplinary knowledge K_Σ is much greater than those of m experts with separated individual disciplinary knowledge K_m, i.e.:

$$K_\Sigma \gg K_m \tag{19}$$

The above lemma can be proven on the basis of the OAR model for internal knowledge representation in the following theorem.

Theorem 8. The *power of multidisciplinary knowledge* states that the *ratio of knowledge space* r_Σ between the knowledge of an expert with coherently m disciplinary knowledge K_Σ and that of a group of m experts with separated individual disciplinary knowledge K_m, i.e.:

$$
\begin{aligned}
r_\Sigma(m, n) &= \frac{K_\Sigma}{K_m} \\
&= \frac{C^2_{m \bullet n}}{\sum\limits_{i=1}^{m} C^2_n} = \frac{\dfrac{(mn)!}{2!(mn-2)!}}{\dfrac{m(n)!}{2!(n-2)!}} \approx \frac{(mn)^2}{mn^2} \\
&= m
\end{aligned}
\tag{20}
$$

where n is the number of *average knowledge objects* or concepts in the discourse disciplines.

Example 3. Reuse the data and context of Example 1, i.e. $m = 12$, $n = 1,000$, Eq. 20 yields that the knowledge space of an expert with coherently 12 disciplinary knowledge is 12 times greater than the sum of that of the 12 experts with separated individual disciplinary knowledge.

Corollary 1. The *ratio of knowledge space* r_1 between the knowledge of an expert with coherently m disciplinary knowledge K_Σ and that of an expert with individual disciplinary knowledge K_1 is:

$$
\begin{aligned}
r_1(m, n) &= \frac{K_\Sigma}{K_1} \\
&= \frac{C^2_{m \bullet n}}{C^2_n} \\
&= m^2
\end{aligned}
\tag{21}
$$

where n is the number of average knowledge objects or concepts in a given discipline.

Corollary 2. The more the interdisciplinary knowledge one acquires, the larger the knowledge space, and hence the higher the possibility for creation and innovation.

Corollary 2 provides a rational answer for a fundamental question in knowledge science: Which his more important in research if there is a need to choose the preference from broadness and depth for individual's knowledge structure? More formally, it can be expressed in the following corollary.

Corollary 3. In knowledge acquisitions toward creativity and problem solving, *broadness* is more important than *depth* in one's knowledge structure.

The above corollary is perfectly in line with the philosophy of holism [3], [12], [18].

6 Conclusions

This paper has explored the theoretical and cognitive foundations of knowledge and knowledge science. As a result, the framework of formal knowledge system (FKS) has been established, where knowledge has been modeled as acquired information in forms of data, behavior, experience, and skills retained in memory through learning. The nature, taxonomy, mathematical models, and manipulation rules of human knowledge have been systematically investigated based on concept algebra. The Hierarchical Abstraction Model (HAM) of knowledge has been developed. The framework of formalized knowledge system has been elaborated based on analyses of formal and empirical knowledge. Then, the theory of knowledge acquisition and the cognitive model of knowledge spaces have been formally described.

The advantages of the hierarchical knowledge architecture in the form of concept networks have been identified as dynamic and evolvable. The advances of the formal knowledge systems have been recognized as their stability and efficiency. It has been proven that in knowledge acquisition and knowledge engineering, broadness is more important than depth. The more the interdisciplinary knowledge one acquires, the larger the knowledge space, and hence the higher the possibility for creation and innovation.

Acknowledgement

The author would like to acknowledge the Natural Science and Engineering Council of Canada (NSERC) for its partial support to this work. The author would like to thank the anonymous reviewers for their valuable comments and suggestions.

References

1. Baddeley, A.: Human Memory: Theory and Practice. Allyn and Bacon, Needham Heights (1990)
2. Debenham, J.K.: Knowledge Systems Design. Prentice Hall, New York (1989)
3. Ellis, D.O., Fred, J.L.: Systems Philosophy. Prentice Hall, Englewood Cliffs (1962)
4. Gabrieli, J.D.E.: Cognitive Neuroscience of Human Memory. Annual Review of Psychology 49, 87–115 (1998)

5. Ganter, B., Wille, R.: Formal Concept Analysis. Springer, Berlin (1999)
6. Gray, P.: Psychology, 2nd edn. Worth Publishers, Inc., New York (1994)
7. Leahey, T.H.: A History of Psychology: Main Currents in Psychological Thought, 4th edn. Prentice-Hall Inc., Upper Saddle River (1997)
8. Matlin, M.W.: Cognition, 4th edn. Harcourt Brace College Publishers, Orlando (1998)
9. Quillian, M.R.: Semantic Memory. In: Minsky, M. (ed.) Semantic Information Processing. MIT Press, Cambridge (1968)
10. Rosenzmeig, M.R., Leiman, A.L., Breedlove, S.M.: Biological Psychology: An Introduction to Behavioral, Cognitive, and Clinical Neuroscience, 2nd edn. Sinauer Associates, Inc., Publishers, Sunderlans (1999)
11. Smith, R.E.: Psychology. West Publishing Co., St. Paul (1993)
12. Sober, E.: Core Questions in Philosophy: A Text with Readings, 2nd edn. Prentice Hall, Englewood Cliffs (1995)
13. Squire, L.R., Knowlton, B., Musen, G.: The Structure and Organization of Memory. Annual Review of Psychology 44, 453–459 (1993)
14. Sternberg, R.J.: In Search of the Human Mind, 2nd edn. Harcourt Brace & Co., Orlando (1998)
15. Wang, Y.: Keynote: On Cognitive Informatics. In: Proc. 1st IEEE International Conference on Cognitive Informatics (ICCI 2002), Calgary, Canada, pp. 34–42. IEEE CS Press, Los Alamitos (2002)
16. Wang, Y.: On Cognitive Informatics, Brain and Mind: A Transdisciplinary. Journal of Neuroscience and Neurophilosophy 4(3), 151–167 (2003)
17. Wang, Y.: Keynote: Cognitive Informatics - Towards the Future Generation Computers that Think and Feel. In: Proc. 5th IEEE International Conference on Cognitive Informatics (ICCI 2006), Beijing, China, pp. 3–7. IEEE CS Press, Los Alamitos (2006)
18. Wang, Y.: Software Engineering Foundations: A Software Science Perspective. CRC Book Series in Software Engineering, vol. II. Auerbach Publications, NY (2007a)
19. Wang, Y.: The Theoretical Framework of Cognitive Informatics. International Journal of Cognitive Informatics and Natural Intelligence 1(1), 1–27 (2007b)
20. Wang, Y.: The OAR Model of Neural Informatics for Internal Knowledge Representation in the Brain. International Journal of Cognitive Informatics and Natural Intelligence 1(3), 64–75 (2007c)
21. Wang, Y.: The Theoretical Framework and Cognitive Process of Learning. In: Proc. 6th International Conference on Cognitive Informatics (ICCI 2007). IEEE CS Press, Los Alamitos (2007d)
22. Wang, Y.: Keynote, On Abstract Intelligence and Its Denotational Mathematics Foundations. In: Proc. 7th IEEE International Conference on Cognitive Informatics (ICCI 2008), Stanford University, CA, USA, pp. 3–13. IEEE CS Press, Los Alamitos (2008a)
23. Wang, Y.: On Contemporary Denotational Mathematics for Computational Intelligence. Transactions of Computational Science 2, 6–29 (2008b)
24. Wang, Y.: On Concept Algebra: A Denotational Mathematical Structure for Knowledge and Software Modeling. International Journal of Cognitive Informatics and Natural Intelligence 2(2), 1–19 (2008c)
25. Wang, Y.: On System Algebra: A Denotational Mathematical Structure for Abstract Systems, Modeling. International Journal of Cognitive Informatics and Natural Intelligence 2(2), 20–43 (2008d)
26. Wang, Y.: On Abstract Intelligence: Toward a Unified Theory of Natural, Artificial, Machinable, and Computational Intelligence. International Journal of Software Science and Computational Intelligence 1(1), 1–17 (2009a)

27. Wang, Y.: On Cognitive Foundations of Creativity and the Cognitive Process of Creation. International Journal of Cognitive Informatics and Natural Intelligence 3(4), 1–16 (2009b)
28. Wang, Y.: Formal Description of the Cognitive Process of Memorization. Transactions of Computational Science 5, 81–98 (2009c)
29. Wang, Y., Kinsner, W., Anderson, J.A., Zhang, D., Yao, Y.Y., Sheu, P., Tsai, J., Pedrycz, W., Latombe, J.-C., Zadeh, L.A., Patel, D., Chan, C.: A Doctrine of Cognitive Informatics. Fundamenta Informaticae 90(3), 1–26 (2009a)
30. Wang, Y., Zadeh, L.A., Yao, Y.: On the System Algebra Foundations for Granular Computing. International Journal of Software Science and Computational Intelligence 1(1), 64–86 (2009b)
31. Wang, Y., Liu, D., Wang, Y.: Discovering the Capacity of Human Memory, Brain and Mind: A Transdisciplinary. Journal of Neuroscience and Neurophilosophy 4(2), 189–198 (2003)
32. Wang, Y., Wang, Y.: Cognitive Informatics Models of the Brain. IEEE Transactions on Systems, Man, and Cybernetics (C) 36(2), 203–207 (2006)
33. Wang, Y., Wang, Y., Patel, S., Patel, D.: A Layered Reference Model of the Brain (LRMB). IEEE Transactions on Systems, Man, and Cybernetics (C) 36(2), 124–133 (2006)
34. Wilson, R.A., Keil, F.C.: The MIT Encyclopedia of the Cognitive Sciences. MIT Press, Cambridge (2001)

On Temporal Properties of Knowledge Base Inconsistency

Du Zhang

Department of Computer Science
California State University
Sacramento, CA 95819-6021

Abstract. Inconsistent knowledge and information in the real world often have to do with not only what conflicting circumstances are but also when they happen. As a result, only scrutinizing the logical forms that cause inconsistency may not be adequate. In this paper we describe our research work on the temporal characteristics of inconsistent knowledge that can exist in an intelligent system. We provide a formal definition for temporal inconsistency that is based on a logical-conflicting and temporal-coinciding dichotomy. The interval temporal logic underpins our treatment of temporally inconsistent propositions in a knowledge base. We also propose a systematic approach to identifying conflicting intervals for temporally inconsistent propositions. The results help delineate the semantic difference between the classical and temporal inconsistency.

Keywords: Interval temporal logic, knowledge inconsistency, temporal inconsistency, interval relations, conflicting intervals.

1 Introduction

The world we live in is full of inconsistency in knowledge and in information. In many circumstances, the conflicting information we encounter in our daily life is inconsequential, causes some minor inconvenience, or generates a few laughter (e.g., the headlines segment in Jay Leno's Tonight Show often includes contradicting but hilarious ads such as "silk handkerchief that is 100% cotton"). However, sometimes the consequence of such inconsistent information can be costly, grievous and devastating. Here are two high-profile cases to wit:

> On September 23, 1999, NASA lost a $125 million Mars Climate Orbiter after its 286-day journey due to the inconsistent navigation information in the English units (pounds force) of measurement sent by the Lockheed Martin engineering team and the NASA's JPL flight control team's calculation in the metric units (newtons) [25]. Since one pound force is equivalent to 4.45 newtons, the result was that "the changes made to the spacecraft's trajectory were actually 4.4 times greater than what the JPL navigation team believed." This caused the space craft to come within 60 km of the Mars,

M.L. Gavrilova et al. (Eds.): Trans. on Comput. Sci. V, LNCS 5540, pp. 20–37, 2009.

*which is about 100 km closer than planned, and about 25 km be-
neath the level at which it could function properly. As a result, the
propulsion system of the orbiter overheated and was subsequently
disabled as it dipped deeply into the atmosphere.*

*On January 2, 2006, there was a coal mine explosion in the Sago
Mine in Sago, West Virginia, which trapped 13 miners. During the
thick of confusion, information was given to the family members
that 12 survivors were found and only one had died (see a news report [29]). This inconsistent information, only to be corrected
shortly after the fact revealing that there was only one survivor and
all the other 12 had perished, caused unimaginable pain and emo-
tional suffering to the families and the public at large.*

In the fields of computer science and artificial intelligence, we also need to come to
grips with inconsistent and conflicting knowledge and information [4, 7, 9, 20, 22].
Software requirement specifications often contain inconsistent information [12]. Inconsistency can also be found in spatial databases [26]. Citation matching problem in
digital libraries [19] and conflicting information on the web [30] are vivid reminders
of how prevalent inconsistent information can be in the real world situations. The
domain knowledge about a particular application that gets codified in an intelligent
computer system's knowledge base (KB) may contain inconsistent and contradictory
information that not only affects the correctness and performance of the system, but
can also have grave consequences. At a time when the information technologies play
such a pivotal and indispensible role in our society, our economy, and our daily life,
we can never underestimate the impact the inconsistent information in computer systems has in our decision making process because people's lives may be at risk. One of
the most widely cited software-related accidents in safety-critical systems involved a
computerized radiation therapy machine called the Therac-25 [21]. Software coding
errors contributed to six known accidents where massive overdoses were involved by
the Therac-25 -- with resultant deaths and serious injuries during the period from June
1985 to January 1987.

When inconsistency in a KB results in two conflicting decisions such as "increase
the radiation dose by 500 units" and "do not increase the radiation dose by 500 units,"
a critical issue is whether the two commands are issued at the same time or coincide
in time. In general for two conflicting propositions, we are interested in knowing not
only that they are inconsistent, but also when they are inconsistent. If the time intervals at which they each hold do not overlap, i.e., they do not occur simultaneously in
the same time period, then there may not be a contradiction. To investigate the temporal properties of inconsistent information that can creep up in a decision making process, we need to resort to a temporal logic to formally establish temporal relationships
between time periods of propositions and to reason about the presence or absence of
temporal inconsistency. There are a number of temporal logic formalisms [1-3, 6, 23].
In this work, we choose to adopt James Allen's interval temporal logic [1-2], which is
based on characterizing actions and events in terms of interval relationships.

Though there have been numerous studies on knowledge base inconsistency
[10-18, 24, 27-28, 31-33] and on temporal logic and its applications [1-3, 6, 23], respectively, the issue of characterizing temporal properties of knowledge base inconsistency in terms of some temporal logic formalism has attracted little attention thus

far as evidenced in the lack of published results in the literature. The focus of this paper is on how to define temporal inconsistency and how to identify conflicting intervals for temporally inconsistent propositions.

The rest of the paper is organized as follows. Section 2 gives a brief overview on the interval temporal logic as was initially defined by James Allen. Section 3 introduces a list of different types of KB inconsistency that will be used as examples and reexamined under the proposed temporal framework. A formal definition of temporal inconsistency is described in Section 4. Section 5 discusses two algorithms that are used to detect conflicting intervals for temporal inconsistency. Section 6 includes a brief account on related work. Finally, Section 7 concludes the paper with remarks on future work.

2 Interval Temporal Logic

Based on the classification in [6], there are several dimensions through which various formalisms of temporal logic can be categorized: propositional vs. first-order, global vs. compositional, branching vs. linear time, points vs. intervals, discrete vs. continuous, and past operators vs. future operators.

James Allen's interval temporal logic (ITL) [1-2] is the formalism we choose to use in this work to delineate temporal inconsistency in a KB. According to Emerson's criteria [6], ITL can be considered as first-order, global, linear time, intervals, continuous and future operators. A brief overview of Allen's interval temporal logic is given in this section. Readers are referred to [1-3, 6, 23] for details.

The fundamental temporal structure in ITL is a linear model of time. ITL starts with one primitive object, the time period or interval, and one primitive relation *Meets*. A time interval represents the time duration of some event that occurs or some property that holds in the world. Two intervals i and j meet if and only if i precedes j, but there is no time between i and j, and i and j do not overlap. There is a set of axioms about *Meets*: (1) every interval has an interval that meets it and another that it meets; (2) intervals can be concatenated to form a larger interval; (3) intervals uniquely define an equivalence class of intervals that meet them; (4) two intervals are equal if both meet the same interval and another interval meets them both; and (5) interval meeting spots can be ordered [2]. In addition, no interval can meet itself. If interval i meets interval j, then j cannot also meet i.

With the primitive object and the primitive relationship in place, a complete list of additional interval relationships can be defined as follows.

$Before(i, j) \equiv_{def} \exists k \ [Meets(i, k) \wedge Meets(k, j)]$

$Overlaps(i, j) \equiv_{def} \exists m \ \exists n \ \exists o \ \exists p \ \exists q \ [Meets(m, i) \wedge Meets(m, n) \wedge Meets(n, j) \wedge$
$\quad\quad Meets(n, p) \wedge Meets(p, q) \wedge Meets(i, q) \wedge Meets(j, o) \wedge Before(i, o)]$

$Equals(i, j) \equiv_{def} \exists m \ \exists n \ [Meets(m, i) \wedge Meets(m, j) \wedge Meets(i, n) \wedge Meets(j, n)]$

$Starts(i, j) \equiv_{def} \exists m \ \exists n \ [Meets(m, i) \wedge Meets(m, j) \wedge Meets(j, n) \wedge Before(i, n)]$

$During(i, j) \equiv_{def} \exists m \ \exists n \ [Meets(m, j) \wedge Meets(j, n) \wedge Before(m, i) \wedge Before(i, n)]$

$Finishes(i, j) \equiv_{def} \exists m \ \exists n \ [Meets(i, n) \wedge Meets(j, n) \wedge Meets(m, j) \wedge Before(m, i)]$

Figure 1 offers a graphic depiction for the relationships. Similar definitions can be given for the inverse relationships of the aforementioned intervals: *MetBy(j, i)*, *After(j, i)*, *OverlappedBy(j, i)*, *StartedBy(j, i)*, *Contains(j, i)*, and *FinishedBy(j, i)*. Hence, there are thirteen interval relationships altogether.

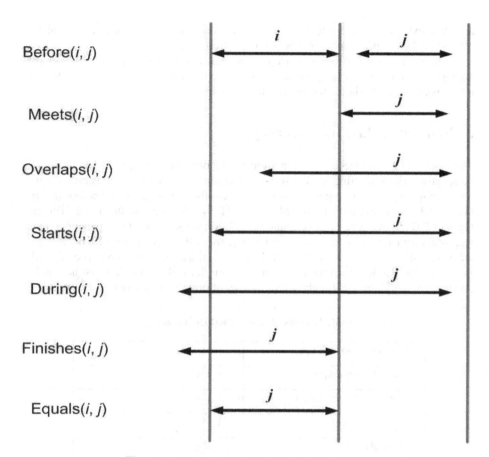

Fig. 1. Interval relationships for actions and events

Definition 1. For a given atomic formula $P(t_1, ..., t_n)$ where P is the predicate symbol and $t_1,...,t_n$, are terms, we introduce a time period to the atom through adding two arguments $\mathscr{S}$, a time point, and $\mathscr{D}$, a duration: $P(t_1, ..., t_n, \mathscr{S}, \mathscr{D})$. The semantic reading of the atom is that the truth value of $P(t_1, ..., t_n)$ is *true* from the point of $\mathscr{S}$ and for the period of $\mathscr{D}$. We can also use $\Delta=[\mathscr{S}, \mathscr{S}+\mathscr{D}]$ to represent the duration and Δ_P to denote the time interval for P.   □

There are two possible interpretations of negation: a strong and a weak interpretation [2]. In the weak interpretation, the negation of a predicate P is true over interval t if and only if P is not true throughout t. In other words, the weak interpretation allows P to be true in some subinterval of t. On the other hand, in the strong interpretation, the negation of a predicate P is true over interval t if and only if P is false throughout t. In this paper, we choose to adopt the strong interpretation in our work. Though the strong interpretation has truth gaps issue in general [2], the application of ITL to characterizing temporal inconsistency will not be affected by it. The reason is that we do not fix an arbitrary time period t and study the conflicting predicates with regard to t. Rather our approach is predicates driven: we identify the conflicting predicates first and then extract the time interval information from them. Thus for any interval we have to deal with, predicates all have persistent truth values.

In the context of this research, we assume that knowledge in a KB observes the *homogeneity* property. A proposition is homogeneous if and only if when it holds for a time period t, it also holds for any sub-period within t [2].

3 Knowledge Base Inconsistency

The issue of knowledge base inconsistency has been largely approached from the perspectives of conflicting logical forms, violations of ontological constraints, epistemic conflicts, lack of complete information, conflicting defaults, defeasible inheritance, and belief revision [10-18, 24, 27-28, 31-34]. A recent summary of different types of KB inconsistency can be found in [34]. Since the focus of this paper is on the temporal characterization of KB inconsistency, we will use the types of KB inconsistency in [34] as examples to illustrate how to establish temporal properties for KB inconsistency. Table 1 summarizes what has been defined in [34]. The list here is by no means a complete one. It only serves as a set of possible examples.

Table 1. Types of possible KB inconsistency

Inconsistency Type	Notation
$i\mathscr{T}_1$: Complementary	$L_i \neq L_j$
$i\mathscr{T}_2$: Mutually exclusive	$L_i \neq L_j$
$i\mathscr{T}_3$: Incompatible	$L_i \not\cong L_j$
$i\mathscr{T}_4$: Anti-subtype	$L_i \not\sqsubseteq L_j$
$i\mathscr{T}_5$: Anti-supertype	$L \not\Leftrightarrow (\sqcup L_k)$
$i\mathscr{T}_6$: Asymmetric	$L_i \not\downarrow L_j$
$i\mathscr{T}_7$: Anti-inverse	$L_i \not\rightleftarrows L_j$
$i\mathscr{T}_8$: Mismatching	$L \not\cong (\sqcap L_k)$
$i\mathscr{T}_9$: Disagreeing	$L_i \not\gtrsim L_j$
$i\mathscr{T}_{10}$: Contradictory	$L_i \neq L_j$
$i\mathscr{T}_{11}$: iProbVal	$Prob(L) \notin Prob(\Omega)$

Here are brief explanations to the aforementioned inconsistency types.

$i\mathscr{T}_1$. Given A_1 and A_2 as syntactically identical atoms (same predicate symbol, same arity, and same terms at corresponding positions), A_1 and $\neg A_2$ (or $\neg A_1$ and A_2) are referred to as *complementary* literals. We denote complementary literals as $L_1 \neq L_2$ where L_1 and L_2 are an atom and its negation. For instance:

$$Can\_fly(\text{fred}) \neq \neg Can\_fly(\text{fred})$$

$i\mathscr{S}_2$. Given two literals L_1 and L_2 that are syntactically different and semantically opposite of each other (the assertion of L_1 (L_2) implies the negation of the other L_2 (L_1)), we call L_1 and L_2 as *mutually exclusive* literals and use $L_1 \neq L_2$ to denote it. For example:

$$Animal(\text{seaCucumber}) \neq Vegetable(\text{seaCucumber})$$

$i\mathscr{S}_3$. For two literals L_1 and L_2 that are syntactically different, but logically equivalent, we call them *synonymous*, and denoted $L_1 \cong L_2$. For instance, $Father(x, \text{john}) \cong Male\_parent(x, \text{john})$. Given L_1 and L_2 that are a complementary pair of synonymous literals, we call them *incompatible* literals, and denote $L_1 \ncong L_2$. Here is an example of incompatible literals:

$$Expensive(x) \ncong \neg High\_priced(x)$$

$i\mathscr{S}_4$. For two literals (or concepts as in description logics) L_1 and L_2, if L_1 is a *subtype* or a *specialization* of L_2, then we say that L_1 is subsumed by L_2 (or L_2 subsumes L_1) and use $L_1 \sqsubseteq L_2$ to denote that. For example, $Surgeon(\text{john}) \sqsubseteq Doctor(\text{john})$. If L_1 is no longer a subtype of L_2, then we call L_1 and L_2 *anti-subtype* literals and use the following to denote it: $L_1 \not\sqsubseteq L_2$. For example:

$$Senator(x) \not\sqsubseteq \neg Legislator(x)$$

$i\mathscr{S}_5$. If L is a *supertype* consisting of a list of subtypes denoted as $\uplus L_i$, we use $L \Leftrightarrow \uplus L_i$ to represent the fact that L corresponds to the subtypes in $\uplus L_i$. When $\uplus L_i$ is no longer a set of subtypes defining the supertype L, we say that $\uplus L_i$ is *anti-supertype* with regard to L and use $L \nLeftrightarrow (\uplus L_i)$ to denote that. For instance, the supertype *Agent* in a particular organization consists of the following subtypes: user agents, broker agents, resource agents and ontology agents.

$$Agent(x) \Leftrightarrow (\uplus L_i)$$

where $\uplus L_i = [User\_agent(x) \vee Broker\_agent(x) \vee Resource\_agent(x)$

$$\vee\ Ontology\_agent(x)]$$

Let $\uplus L_i'$ be $[User\_agent(x) \vee Broker\_agent(x)]$, then $Agent(x) \nLeftrightarrow (\uplus L_i')$

$i\mathscr{S}_6$. Given two literals L_1 and L_2 that share the same predicate, if the predicate in L_1 and L_2 is a *symmetric* relation (i.e., if L_1 is $p(x, y)$ and L_2 is $p(y, x)$, then $\forall x\ \forall y\ [p(x, y) \supset p(y, x)]$), L_1 and L_2 are referred to as *symmetric* and are denoted as $L_1 \curlyvee L_2$. When the predicate in L_1 and L_2 is a symmetric relation, but L_1 and L_2 are no longer symmetric literals, we say that L_1 and L_2 are *asymmetric* and use $L_1 \not\curlyvee L_2$ to denote that. For instance,

$$Married\_to(\text{john}, \text{jane}) \not\curlyvee Married\_to(\text{jane}, \text{mike})$$

$i\mathscr{S}_7$. When two predicates represent relationships that are inverse of each other, we call them *inverse* predicates. Literals L_1 and L_2, are referred to as inverse literals when they contain inverse predicates and represent inverted relationships. We use $L_1 \asymp L_2$ to denote that. When predicates in L_1 and L_2 represent inverse relationships but L_1 and

L_2 are no longer inverse literals, we say that L_1 and L_2 are *anti-inverse* and denote it with $L_1 \neq L_2$. For instance:

$$Send\_msg\_to(\text{agent1, agent2}) \neq Received\_msg\_from(\text{agent2, agent3})$$

*iℱ*₈. When a compound predicate (L) is fully defined through a logical expression of other predicates ($\mathfrak{m} L_i$), we use $L \equiv \mathfrak{m} L_i$ to denote it. When a logical expression $\mathfrak{m} L_i'$ is derived for the compound predicate L that is fully defined by $L \equiv \mathfrak{m} L_i$, if $(\mathfrak{m} L_i') \not\equiv (\mathfrak{m} L_i)$, we say that L and $(\mathfrak{m} L_i')$ are *mismatching* and use $L \not\equiv (\mathfrak{m} L_i')$ to denote that. For instance,

$$Mobile\_agent(\text{agent1}) \equiv [Executing(\text{agent1, host1}) \wedge Executing(\text{agent1, host2})$$

$$\wedge \text{ host1} \neq \text{host2}]$$

$$Mobile\_agent(\text{agent1}) \not\equiv [Executing(\text{agent1, host1})]$$

*iℱ*₉. Given L_1 and L_2 for the same proposition and L_2 is at a more concrete level of abstraction than L_1, we call them *reified* literals and denote it with $L_1 \geqq L_2$. If reified quantities in L_1 and L_2 are no longer compatible, we say that L_1 and L_2 are *disagreeing* and use $L_1 \gneqq L_2$ to denote it. For example:

$$Memory\_capacity(\text{agent1, 2GB}) \gneqq Memory\_capacity(\text{agent1, 1500MB})$$

*iℱ*₁₀. Given literals L_1 and L_2 with either the same or different predicate symbols, if they contain attributes (terms) which violate type restrictions or integrity constraints, we refer to L_1 and L_2 as *contradictory* and denote it with $L_1 \neq L_2$. For instance:

$$Birth\_date(\text{john, 3-3-1960}) \neq Hire\_date(\text{john, 5-3-1959})$$

*iℱ*₁₁. Given a set Ω of sentences and their probabilities, and a literal L that logically follows from Ω, if Prob(L) is outside the convex region specified by the extreme values of probabilities for Δ and L, then we say that the assignment of the probabilistic truth value for L is inconsistent [9]. We call this type of inconsistency as *inconsistent probabilistic truth value (iProbVal)* and use $\text{Prob}(L) \notin \text{Prob}(\Omega)$ to denote it.

To facilitate the temporal characterization of the aforementioned inconsistency, we assume that all the predicates include the two temporal arguments as specified in Definition 1.

4 Temporal Inconsistency

In this section, we provide a formal delineation of the underpinnings for temporal inconsistency. We first introduce the concept of a coinciding period between two formulas (or between a formula and a logic expression). We then give a formal definition on temporal inconsistency. Finally we define the types of temporal inconsistency.

4.1 Definition of Temporal Inconsistency

Given two formulas P and Q, the time periods of P and Q are said to be *coinciding*, denoted as $cP(\Delta_P, \Delta_Q)$, when one of the following interval relationships holds between the two: *Overlaps, OverlappedBy, Starts, StartedBy, During, Contains, Finishes, FinishedBy,* or *Equals*.

$cP(\Delta_P, \Delta_Q) \equiv_{def} [Overlaps(\Delta_P, \Delta_Q) \vee Starts(\Delta_P, \Delta_Q) \vee During(\Delta_P, \Delta_Q)$
$\qquad \vee Finishes(\Delta_P, \Delta_Q) \vee Equals(\Delta_P, \Delta_Q) \vee OverlappedBy(\Delta_Q, \Delta_P)$
$\qquad \vee StartedBy(\Delta_Q, \Delta_P) \vee Contains(\Delta_Q, \Delta_P) \vee FinishedBy(\Delta_Q, \Delta_P)]$

We use $\neg cP(\Delta_P, \Delta_Q)$ to represent the fact that P and Q are non-coinciding formulas.

Let $\mathscr{L}$ be a logic expression in $\{ \uplus L_k, \, \Cap L_k, \, \Omega \}$. Let $\Delta_{\mathscr{L}}$ denote the time interval shared by literals in $\uplus L_k$ or $\Cap L_k$ or Ω.

$$\Delta_{\mathscr{L}} = \{ \cap \Delta_L \mid (L \in \uplus L_k) \vee (L \in \Cap L_k) \vee (L \in \Omega) \}$$

The time periods of a formula P and a logic expression $\mathscr{L}$ are *coinciding*, denoted similarly as $cP(\Delta_P, \Delta_{\mathscr{L}})$, when one of the following interval relationships holds:

$cP(\Delta_P, \Delta_{\mathscr{L}}) \equiv_{def} [Overlaps(\Delta_P, \Delta_{\mathscr{L}}) \vee Starts(\Delta_P, \Delta_{\mathscr{L}}) \vee During(\Delta_P, \Delta_{\mathscr{L}})$
$\qquad \vee Finishes(\Delta_P, \Delta_{\mathscr{L}}) \vee Equals(\Delta_P, \Delta_{\mathscr{L}}) \vee OverlappedBy(\Delta_{\mathscr{L}}, \Delta_P)$
$\qquad \vee StartedBy(\Delta_{\mathscr{L}}, \Delta_P) \vee Contains(\Delta_{\mathscr{L}}, \Delta_P) \vee FinishedBy(\Delta_{\mathscr{L}}, \Delta_P)]$

$\neg cP(\Delta_P, \Delta_Q)$ is true when P and $\mathscr{L}$ do not have a coinciding interval.

For two literals L_i and L_j *consistent* with each other, we use $\models (L_i, L_j)$ to denote that. We say that L_i and L_j are *inconsistent* with each other, denoted as $\nvDash (L_i, L_j)$, if

$$\nvDash (L_i, L_j) \equiv_{def} [(L_i \neq L_j) \vee (L_i \neq L_j) \vee (L_i \ngeqslant L_j) \vee (L_i \nsubseteq L_j) \vee (L_i \not\Supset L_j) \vee (L_i \neq L_j)$$
$$\vee (L_i \ngeqslant L_j) \vee (L_i \neq L_j)].$$

When there is a *consistent* circumstance involving a literal L_i and a logic expression $\mathscr{L}$ in either of the following cases,

$$(L_i \Leftrightarrow \uplus L_k) \vee (L_i \equiv \Cap L_k) \vee (Prob(L_i) \in Prob(\Omega))$$

we use $\models (L_i, \mathscr{L})$ to denote it. On the other hand, when there is an *inconsistent* circumstance involving the literal L_i and its related logic expression $\mathscr{L}$ in either of the following cases:

$$\nvDash (L_i, \mathscr{L}) \equiv_{def} [(L_i \nLeftrightarrow \uplus L_k) \vee (L_i \not\equiv \Cap L_k) \vee (Prob(L_i) \notin Prob(\Omega))]$$

we use $\nvDash (L_i, \mathscr{L})$ to denote it.

Now, we are in a position to accurately define what temporal inconsistency entails.

Definition 2. Given two literals L_i and L_j, or a literal L_i and its related logical expression $\mathscr{L}$, *temporal inconsistency*, denoted as $\nvDash t$, between L_i and L_j, or between L_i and $\mathscr{L}$, can be formally defined as follows:

$$\nvDash t(L_i, L_j) \equiv_{def} [\nvDash (L_i, L_j) \wedge cP(\Delta_{Li}, \Delta_{Lj})],$$
$$\nvDash t(L_i, \mathscr{L}) \equiv_{def} [\nvDash (L_i, \mathscr{L}) \wedge cP(\Delta_{Li}, \Delta_{\mathscr{L}})]$$

Combining the two, we have

$$\nvDash t(L_i, (L_j \vee \mathscr{L})) \equiv_{def} [\nvDash (L_i, L_j) \wedge cP(\Delta_{Li}, \Delta_{Lj})] \vee [\nvDash (L_i, \mathscr{L})] \wedge cP(\Delta_{Li}, \Delta_{\mathscr{L}})]. \quad \square$$

Thus, L_i and L_j, or L_i and $\mathscr{L}$, are said to be temporally inconsistent if L_i and L_j, or L_i and $\mathscr{L}$, are logically conflicting and their time intervals are temporally coinciding.

Let $\varDelta = (\Delta_{Li} \cap \Delta_{Lj}) \vee (\Delta_{Li} \cap \Delta_{\mathscr{L}})$ denote the largest sub-interval that is contained in Δ_{Li} and Δ_{Lj}, or in Δ_{Li} and $\Delta_{\mathscr{L}}$. Thus, $\varDelta$ is the maximum coinciding interval in either $cP(\Delta_{Li}, \Delta_{Lj})$ or $cP(\Delta_{Li}, \Delta_{\mathscr{L}})$. For $\nvDash \iota(L_i, L_j)$ or $\nvDash \iota(L_i, \mathscr{L})$, when we have to be specific about the interval over which temporal inconsistency takes place, we use the following notation to indicate that L_i and L_j, or L_i and $\mathscr{L}$, are temporally inconsistent over the maximum coinciding period $\varDelta$:

$$\nvDash \iota(L_i, L_j)_\varDelta, \quad \nvDash \iota(L_i, \mathscr{L})_\varDelta.$$

Temporal consistency, denoted as $\vDash \iota(L_i, L_j)$, or $\vDash \iota(L_i, \mathscr{L})$, can be defined as

$$\vDash \iota(L_i, L_j) \equiv_{def} [(\vDash (L_i, L_j) \wedge Equals(\Delta_{Li}, \Delta_{Lj})) \vee (\nvDash (L_i, L_j) \wedge \neg cP(\Delta_{Li}, \Delta_{Lj}))],$$
$$\vDash \iota(L_i, \mathscr{L}) \equiv_{def} [(\vDash (L_i, \mathscr{L}) \wedge Equals(\Delta_{Li}, \Delta_{\mathscr{L}})) \vee (\nvDash (L_i, \mathscr{L}) \wedge \neg cP(\Delta_{Li}, \Delta_{\mathscr{L}}))].$$

Hence, two literals L_i and L_j (or a literal L_i and its related logic expression $\mathscr{L}$) are temporally consistent if they are consistent and have the same time interval[1], or if they are inconsistent but non-coinciding. Similarly, we can use $\vDash \iota(L_i, L_j)_\varDelta$ (or $\vDash \iota(L_i, \mathscr{L})_\varDelta$) to denote that L_i and L_j (or L_i and $\mathscr{L}$) are temporally consistent over $\varDelta$[2].

In a nutshell, temporal inconsistency hinges on a logical-temporal dichotomy as shown below in Table 2.

Table 2. Logical-temporal dichotomy for temporal inconsistency

Logical Dimension / Temporal Dimension	Inconsistent $\nvDash (L_i, L_j)$ or $\nvDash (L_i, \mathscr{L})$	Consistent $\vDash (L_i, L_j)$ or $\vDash (L_i, \mathscr{L})$
Coinciding $cP(\Delta_{Li}, \Delta_{Lj})$ or $cP(\Delta_{Li}, \Delta_{\mathscr{L}})$	$\nvDash \iota(L_i, L_j)/\nvDash \iota(L_i, \mathscr{L})$	$\vDash \iota(L_i, L_j)/\vDash \iota(L_i, \mathscr{L})$
Non-coinciding $\neg cP(\Delta_{Li}, \Delta_{Lj})$ or $\neg cP(\Delta_{Li}, \Delta_{\mathscr{L}})$	$\vDash \iota(L_i, L_j)/\vDash \iota(L_i, \mathscr{L})$	$\vDash \iota(L_i, L_j)/\vDash \iota(L_i, \mathscr{L})$

4.2 Types of Temporal Inconsistency

For $\nvDash \iota(L_i, L_j)$ or $\nvDash \iota(L_i, \mathscr{L})$, depending on the interval relationship between the two literals, or the literal and its related logical expression, we have the following types of temporal inconsistency:

---

[1] The reason that we insist on the consistent literals having the same time interval is due to the fact that any other interval relationship between the two would result in them having opposite truth values during a sub-interval in either Δ_{Li} or Δ_{Lj}.

[2] When $\neg cP(\Delta_{Li}, \Delta_{Lj})$, then $\varDelta$ in $\vDash \iota(L_i, L_j)_\varDelta$ may represent two disjoint time intervals.

(1). *Congruent*: denoted as $\nvdash_t(L_i, L_j)^c{}_\Delta$, if

  $[\nvdash(L_i, L_j) \wedge Equals(\Delta_{Li}, \Delta_{Lj})]$;
  or denoted as $\nvdash_t(L_i, \mathscr{L})^c{}_\Delta$, if
  $[\nvdash(L_i, \mathscr{L}) \wedge Equals(\Delta_{Li}, \Delta_{\mathscr{L}})]$.

(2). *Subsuming*: denoted as $\nvdash_t(L_i, L_j)^s{}_\Delta$, if

  $[\nvdash(L_i, L_j) \wedge [Starts(\Delta_{Li}, \Delta_{Lj}) \vee During(\Delta_{Li}, \Delta_{Lj}) \vee Finishes(\Delta_{Li}, \Delta_{Lj})$
  $\vee StartedBy(\Delta_{Lj}, \Delta_{Li}) \vee Contains(\Delta_{Lj}, \Delta_{Li}) \vee FinishedBy(\Delta_{Lj}, \Delta_{Li})]]$;
  or denoted as $\nvdash_t(L_i, \mathscr{L})^s{}_\Delta$, if
  $[\nvdash(L_i, \mathscr{L}) \wedge [Starts(\Delta_{Li}, \Delta_{\mathscr{L}}) \vee During(\Delta_{Li}, \Delta_{\mathscr{L}}) \vee Finishes(\Delta_{Li}, \Delta_{\mathscr{L}})$
  $\vee StartedBy(\Delta_{\mathscr{L}}, \Delta_{Li}) \vee Contains(\Delta_{\mathscr{L}}, \Delta_{Li}) \vee FinishedBy(\Delta_{\mathscr{L}}, \Delta_{Li})]]$.

(3). *Overlapping*: denoted as $\nvdash_t(L_i, L_j)^o{}_\Delta$, if

  $[\nvdash(L_i, L_j) \wedge [Overlaps(\Delta_{Li}, \Delta_{Lj}) \vee OverlappedBy(\Delta_{Lj}, \Delta_{Li})]]$;
  or denoted as $\nvdash_t(L_i, L_j)^o{}_\Delta$, if
  $[\nvdash(L_i, \mathscr{L}) \wedge [Overlaps(\Delta_{Li}, \Delta_{\mathscr{L}}) \vee OverlappedBy(\Delta_{\mathscr{L}}, \Delta_{Li})]]$.

For the congruent case, temporal inconsistency for both literals will be *persistent*. For the subsuming case, the literal (or the expression) whose interval is subsumed by that of the other will have a persistent temporal inconsistency. In the overlapping case, there is a chance for both literals (or the literal and its related expression) to be still temporally consistent.

Given a set of time intervals $\{\Delta_i, ..., \Delta_k\}$, we use the following to indicate the concatenation of intervals in the set in the chronological order:

$$\bigwedge \{\Delta_i, ..., \Delta_k\}.$$

Depending on the given intervals, there are different scenarios for the outcomes of the concatenation. In this paper, we only consider the case in which the result is a continuous interval that contains no gap.

Based on the concepts of congruent, subsuming and overlapping temporal inconsistency, we can define the following properties for a given literal.

For a set Ω of literals and $L_i \in \Omega$,

(1). L_i is *completely temporal inconsistent*, denoted as $\mathsf{CTI}(L_i)$, if either of the following holds:

  - $\exists L_j \in \Omega \ [\nvdash_t(L_i, L_j)^c{}_\Delta \vee \nvdash_t(L_i, L_j)^s{}_\Delta]$
  - $\exists \mathscr{L} \in \Omega \ [\nvdash_t(L_i, \mathscr{L})^c{}_\Delta \vee \nvdash_t(L_i, \mathscr{L})^s{}_\Delta]$
  - $\exists L_j,..,L_k \in \Omega \ [(\nvdash(L_i, L_j) \wedge ... \wedge \nvdash(L_i, L_k)) \wedge (\nvdash_t(L_i, L_j)^o{}_{\Delta'} \wedge ...$
      $\wedge \nvdash_t(L_i, L_k)^o{}_{\Delta''}) \wedge (\Delta_{Li} = \bigwedge \{(\Delta_{Li} \cap \Delta_{Lj}), ..., (\Delta_{Li} \cap \Delta_{Lk})\})]$.

(2). L_i is *partially temporal inconsistent*, denoted as $\mathsf{PTI}(L_i)$, if the following holds:

  $$\exists L_j \in \Omega \ \exists \Delta' \subset \Delta_{Li} \ \neg \exists L_k \in \Omega \ [\nvdash_t(L_i, L_j) \wedge \nvdash_t(L_i, L_k)_{\Delta'}].$$

(3). L_i is *fully temporal consistent*, denoted as $\mathsf{FTC}(L_i)$, if the following holds:

  $$\forall L_j \in \Omega \ [\vDash_t(L_i, L_j)]$$

5 Detecting Temporal Inconsistency

There are ways to identify logically conflicting literals in a KB. The work in [33], for instance, describes a fixpoint semantics based approach. Once conflicting literals are identified in a KB, we need to ascertain whether their time intervals are coinciding to determine if they constitute temporal inconsistency. If two literals (or a literal and its related logic expression) are temporally inconsistent, we want to identify the coinciding or conflict interval between the two. Algorithms 1 (for literal-literal) and 2 (for literal-expression) define the conditions to be used to identify specific temporal relations and the resultant conflict intervals with regard to different temporal relationships. Once we obtain the conflict intervals for all temporal inconsistent cases in a KB, we are in a position to characterize the temporal properties for the entire KB.

Given two literals L_i and L_j with their respective time intervals Δ_{Li}, Δ_{Lj}, if $\nvDash (L_i, L_j)$ and $cP(\Delta_{Li}, \Delta_{Lj})$, then their conflicting interval can be obtained according to Algorithm 1.

Algorithm 1. Temporal Inconsistency Detection (literal-literal): $TID_{LL}(.)$

Input: Literals L_i, L_j, and their time intervals Δ_{Li}, Δ_{Lj}, respectively. The intervals can be expressed as follows: $\Delta_{Li} = [\mathscr{S}_i, \mathscr{S}_i + \mathscr{D}_i]$, $\Delta_{Lj} = [\mathscr{S}_j, \mathscr{S}_j + \mathscr{D}_j]$.

Output: Temporal relation between the two literals, and a determination on $\nvDash_t(L_i, L_j)$. If $\nvDash_t(L_i, L_j)$ or $\nvDash_t(L_j, L_i)$, the algorithm also returns the conflicting interval:

  return(*temporal_relation, determination, conflict_interval*).

Initialization {
    temporal_relation = ' ';
    determination = 0;   //A flag that indicates if a given pair of literals is
                               //temporally inconsistent.
                               //0: ($\vDash_t(L_i, L_j) \vee \vDash_t(L_j, L_i)$);
                               //1: $\nvDash_t(L_i, L_j)$;
                               //2: $\nvDash_t(L_j, L_i)$.
    conflict_interval = ∅;   //conflict interval for L_i and L_j that are $\nvDash_t(L_i, L_j)$
                               //or $\nvDash_t(L_j, L_i)$.
}

if $((\mathscr{S}_i + \mathscr{D}_i) < \mathscr{S}_j)$
    then { *temporal_relation* = '*Before*(Δ_{Li}, Δ_{Lj})';
        determination = 0;
        conflict_interval = ∅;}
else if $((\mathscr{S}_j + \mathscr{D}_j) < \mathscr{S}_i)$
then { *temporal_relation* = '*After*(Δ_{Lj}, Δ_{Li})';
    determination = 0;
    conflict_interval = ∅;}
else if $((\mathscr{S}_i + \mathscr{D}_i) = \mathscr{S}_j)$
then { *temporal_relation* = '*Meets*(Δ_{Li}, Δ_{Lj})';
    determination = 0;
    conflict_interval = ∅;}

else if $((\mathscr{S}_j + \mathscr{D}_j) = \mathscr{S}_i)$
then { *temporal_relation* = '*MetBy*(Δ_{Lj}, Δ_{Li})';
    determination = 0;
    conflict_interval = $\varnothing$;}
else if $((\mathscr{S}_i=\mathscr{S}_j) \wedge (\mathscr{D}_i+ k = \mathscr{D}_j) \wedge (0 < k))$
then { *temporal_relation* = '*Starts*(Δ_{Li}, Δ_{Lj})';
    determination = 1;
    conflict_interval = Δ_{Li};}
else if $((\mathscr{S}_j=\mathscr{S}_i) \wedge (\mathscr{D}_j+ k = \mathscr{D}_i) \wedge (0 < k))$
then { *temporal_relation* = '*StartedBy*(Δ_{Lj}, Δ_{Li})';
    determination = 2;
    conflict_interval = Δ_{Lj};}
else if $((\mathscr{S}_j < \mathscr{S}_i) \wedge (\mathscr{D}_i < \mathscr{D}_j))$
then { *temporal_relation* = '*During*(Δ_{Li}, Δ_{Lj})';
    determination = 1;
    conflict_interval = Δ_{Li};}
else if $((\mathscr{S}_i < \mathscr{S}_j) \wedge (\mathscr{D}_j < \mathscr{D}_i))$
then { *temporal_relation* = '*Contains*(Δ_{Lj}, Δ_{Li})';
    determination = 2;
    conflict_interval = Δ_{Lj};}
else if $((\mathscr{S}_i=\mathscr{S}_j+k) \wedge (\mathscr{D}_i+ k = \mathscr{D}_j) \wedge (0 < k))$
then { *temporal_relation* = '*Finishes*(Δ_{Li}, Δ_{Lj})';
    determination = 1;
    conflict_interval = Δ_{Li};}
else if $((\mathscr{S}_j=\mathscr{S}_i+k) \wedge (\mathscr{D}_j+ k = \mathscr{D}_i) \wedge (0 < k))$
then { *temporal_relation* = '*FinishedBy*(Δ_{Lj}, Δ_{Li})';
    determination = 2;
    conflict_interval = Δ_{Lj};}
else if $((\mathscr{S}_j=\mathscr{S}_i+k) \wedge (\mathscr{S}_j < \mathscr{S}_i+ \mathscr{D}_i) \wedge ((\mathscr{S}_i+ \mathscr{D}_i) < (\mathscr{S}_j + \mathscr{D}_j)) \wedge (0 < k < \mathscr{D}_i))$
    then { *temporal_relation* = '*Overlaps*(Δ_{Li}, Δ_{Lj})';
        determination = 1;
        conflict_interval = $[\mathscr{S}_i+ k, \mathscr{S}_i+ \mathscr{D}_i]$;}
else if $((\mathscr{S}_i=\mathscr{S}_j+k) \wedge (\mathscr{S}_i < \mathscr{S}_j+ \mathscr{D}_j) \wedge ((\mathscr{S}_j+ \mathscr{D}_j) < (\mathscr{S}_i + \mathscr{D}_i)) \wedge (0 < k < \mathscr{D}_j))$
    then { *temporal_relation* = '*OverlappedBy*(Δ_{Lj}, Δ_{Li})';
        determination = 2;
        conflict_interval = $[\mathscr{S}_j+ k, \mathscr{S}_j+ \mathscr{D}_j]$;}
else if $((\mathscr{S}_i=\mathscr{S}_j) \wedge (\mathscr{D}_i = \mathscr{D}_j))$
then { *temporal_relation* = '*Equals*(Δ_{Li}, Δ_{Lj})';
    determination = 1;
    conflict_interval = $(\Delta_{Li} \vee \Delta_{Lj})$;}

return(*temporal_relation, determination, conflict_interval*);   □

Given a literal L_i and a logical expression $\mathscr{L}$ with their respective time intervals Δ_{Li}, $\Delta_{\mathscr{L}}$, if $\nvDash (L_i, \mathscr{L})$ and $cP(\Delta_{Li}, \Delta_{\mathscr{L}})$, then their conflicting interval can be obtained according to Algorithm 2.

Algorithm 2. Temporal Inconsistency Detection (literal-expression): $TID_{LE}(.)$

Input: A literal L_i, its related logical expression $\mathscr{L}$, and their time intervals Δ_{Li}, $\Delta_{\mathscr{L}}$, respectively. The intervals can be expressed as follows:
$$\Delta_{Li} = [\mathscr{S}_i, \mathscr{S}_i + \mathscr{D}_i], \Delta_{\mathscr{L}} = [\mathscr{S}_{\mathscr{L}}, \mathscr{S}_{\mathscr{L}} + \mathscr{D}_{\mathscr{L}}].$$

Output: Temporal relation between the literal and the expression, and a determination on $\nvDash_t(L_i, \mathscr{L})$. If $\nvDash_t(L_i, \mathscr{L})$ or $\nvDash_t(\mathscr{L}, L_i)$, the algorithm also returns the conflicting interval:
           (**return**(*temporal_relation, determination, conflict_interval*).

Initialization {
     temporal_relation = ' ';
     determination = 0;   //A flag that indicates if a literal and its related
                                                //logical expression is temporally inconsistent.
                                                //0: $(\vDash_t(L_i, \mathscr{L}) \vee \vDash_t(\mathscr{L}, L_i))$;
                                                //1: $\nvDash_t(L_i, \mathscr{L})$;
                                                //2: $\nvDash_t(\mathscr{L}, L_i)$.
     conflict_interval = $\varnothing$;   //conflict interval for L_i and $\mathscr{L}$ that are $\nvDash_t(L_i, \mathscr{L})$ or
                                                // $\nvDash_t(\mathscr{L}, L_i)$.

}
if $((\mathscr{S}_i + \mathscr{D}_i) < \mathscr{S}_{\mathscr{L}})$
        then { *temporal_relation* = '*Before*(Δ_{Li}, $\Delta_{\mathscr{L}}$)';
                 determination = 0;
                 conflict_interval = $\varnothing$;}
else if $((\mathscr{S}_{\mathscr{L}} + \mathscr{D}_{\mathscr{L}}) < \mathscr{S}_i)$
then { *temporal_relation* = '*After*($\Delta_{\mathscr{L}}$, Δ_{Li})';
         determination = 0;
         conflict_interval = $\varnothing$;}
else if $((\mathscr{S}_i + \mathscr{D}_i) = \mathscr{S}_j)$
        then { *temporal_relation* = '*Meets*(Δ_{Li}, $\Delta_{\mathscr{L}}$)';
                 determination = 0;
                 conflict_interval = $\varnothing$;}
else if $((\mathscr{S}_{\mathscr{L}} + \mathscr{D}_{\mathscr{L}}) = \mathscr{S}_i)$
then { *temporal_relation* = '*MetBy*($\Delta_{\mathscr{L}}$, Δ_{Li})';
         determination = 0;
         conflict_interval = $\varnothing$;}
else if $((\mathscr{S}_i = \mathscr{S}_{\mathscr{L}}) \wedge (\mathscr{D}_i + k = \mathscr{D}_{\mathscr{L}}) \wedge (0 < k))$
then { *temporal_relation* = '*Starts*(Δ_{Li}, $\Delta_{\mathscr{L}}$)';
         determination = 1;
         conflict_interval = Δ_{Li};}
else if $((\mathscr{S}_{\mathscr{L}} = \mathscr{S}_i) \wedge (\mathscr{D}_{\mathscr{L}} + k = \mathscr{D}_i) \wedge (0 < k))$
then { *temporal_relation* = '*StartedBy*($\Delta_{\mathscr{L}}$, Δ_{Li})';
         determination = 2;
         conflict_interval = $\Delta_{\mathscr{L}}$;}
else if $((\mathscr{S}_{\mathscr{L}} < \mathscr{S}_i) \wedge (\mathscr{D}_i < \mathscr{D}_{\mathscr{L}}))$
then { *temporal_relation* = '*During*(Δ_{Li}, $\Delta_{\mathscr{L}}$)';

```
      determination = 1;
      conflict_interval = Δ_Li;}
else if ((𝒮_i < 𝒮_𝓛) ∧ (𝒟_𝓛 < 𝒟_i))
then { temporal_relation = 'Contains(Δ_𝓛, Δ_Li)';
      determination = 2;
      conflict_interval = Δ_𝓛;}
else if ((𝒮_i=𝒮_𝓛 + k) ∧ (𝒟_i+ k = 𝒟_𝓛) ∧ (0 < k))
then { temporal_relation = 'Finishes(Δ_Li, Δ_𝓛)';
      determination = 1;
      conflict_interval = Δ_Li;}
else if ((𝒮_𝓛 = 𝒮_i+k) ∧ (𝒟_𝓛 + k = 𝒟_i) ∧ (0 < k))
then { temporal_relation = 'FinishedBy(Δ_𝓛, Δ_Li)';
      determination = 2;
      conflict_interval = Δ_𝓛;}
else if ((𝒮_𝓛 = 𝒮_i+k) ∧ (𝒮_𝓛 < 𝒮_i+ 𝒟_i) ∧ ((𝒮_i+ 𝒟_i) < (𝒮_𝓛 + 𝒟_𝓛)) ∧ (0 < k < 𝒟_i))
then { temporal_relation = 'Overlaps(Δ_Li, Δ_𝓛)';
      determination = 1;
      conflict_interval = [𝒮_i+ k, 𝒮_i+ 𝒟_i];}
else if ((𝒮_i=𝒮_𝓛 + k) ∧ (𝒮_i < 𝒮_𝓛 + 𝒟_𝓛) ∧ ((𝒮_𝓛 + 𝒟_𝓛) < (𝒮_i + 𝒟_i)) ∧ (0 < k < 𝒟_𝓛))
then { temporal_relation = 'OverlappedBy(Δ_𝓛, Δ_Li)';
      determination = 2;
      conflict_interval = [𝒮_𝓛 + k, 𝒮_𝓛 + 𝒟_𝓛];}
else if ((𝒮_i=𝒮_𝓛) ∧ (𝒟_i = 𝒟_𝓛))
then { temporal_relation = 'Equals(Δ_Li, Δ_𝓛)';
      determination = 1;
      conflict_interval = (Δ_Li ∨ Δ_𝓛);}
```

return(*temporal_relation, determination, conflict_interval*);   ☐

Example 1. If a KB yields the following pair of literals with the last two arguments indicating the $\mathscr{S}$ and $\mathscr{D}$ values, respectively:

L_1: *Memory_capacity(agent1,* 2GB, **3, 12**);
L_2: *Memory_capacity(agent1,* 1500MB, **5, 20**)

then this conflict case is of $i\mathscr{S}_9$. Applying **TID$_{LL}$**(L_1, L_2, [**3, 15**], [**5, 25**]), we obtain the following:

temporal_relation: *Overlaps*(Δ_{L1}, Δ_{L2});
determination: $\nvDash_t(L_1, L_2)$;
conflict_interval: [**5, 15**].

This is an *overlapping* temporal inconsistency ($\nvDash_t(L_1, L_2)^o_{[5,15]}$). If there is no other literals in the KB that conflict with either of L_1 and L_2, then we have PTI(L_1) and PTI(L_2).

Example 2. If we have the following literals where host2 ≠ host3

L_3: *Deployed*(agent1, host2, **3, 12**);
L_4: *Deployed*(agent1, host3, **3, 15**)

this is a case of $i\mathcal{F}_2$. Using $\mathbf{TID}_{LL}(L_3, L_4, [\mathbf{3}, \mathbf{15}], [\mathbf{3}, \mathbf{18}])$, we obtain the results below:

> temporal_relation: $Starts(\Delta_{L3}, \Delta_{L4})$;
> determination: $\nvDash_t(L_3, L_4)$;
> conflict_interval: $[\mathbf{3}, \mathbf{15}]$.

It is a *subsuming* temporal inconsistency ($\nvDash_t(L_3, L_4)^s_{[3,15]}$). We have $CTI(L_3)$ and $PTI(L_4)$.

Example 3. In [8], an airline overbooking example was given to argue for the desirability of inconsistency in certain circumstances. Airlines are often engaged in a practice where more tickets are sold for a flight than there are seats on the flight, thus resulting in an inconsistency with the safety regulations (see the constraints below). This is based on the belief by the airlines that there might be enough no shows by the boarding time to the point that the inconsistency will resolve itself. This "overbooking inconsistency" proves to be useful and desirable to the airlines' bottom-line. In the event that this overbooking inconsistency does not resolve itself during the boarding time, it will trigger a number of possible airline actions ranging from upgrading yet-to-be-seated passengers, arranging for a later flight with rewards, to making alternative travel. Following is the set of two formulas in [8] depicting the overbooking inconsistency.

$$[Passenger(p, f, d) \wedge Checkedin(c, f, d) \wedge (Size(c)=x) \wedge (Capacity(f)= y) \wedge (x > y)]$$
$$\supset Overbooked(f, d)$$
$$Commercial(f) \wedge Legal(f) \supset \neg Overbooked(f, d)$$

where p, f, d, c represent list of passengers, flight number, flight date, and checked in passengers, respectively.

Here $\{Overbooked(f, d), \neg Overbooked(f, d)\}$ gives rise to an $i\mathcal{F}_1$ type of inconsistency. Does this also result in temporal inconsistency? If so, exactly when and for how long? With the following timeline in Figure 2, we can use the proposed approach to analyze exactly when temporal inconsistency arises as a result of having a pair of complementary literals whose time intervals overlap.

Fig. 2. Timeline for overbooked event

Let us augment the *Overbooked* and $\neg Overbooked$ predicates with appropriate time intervals. For the safety constraint, $\neg Overbooked$ should have the following time interval:

$$L_5: \neg Overbooked(f, d, \mathbf{t_4}, (\mathbf{t_5} - \mathbf{t_4}))$$

For *Overbooked*, its truth value is true when tickets for the flight are oversold (rather than at the check-in time as was defined in [8]). Even though $Overbooked(f, d, \mathbf{t_2}, (\mathbf{t_4} - \mathbf{t_2}))$ is true, there is no violation of the safety constraint being committed (there is the interval relationship of $Meets(\Delta_{Overbooked}, \Delta_{\neg Overbooked})$ here). However, it will raise the flag to the airline that there is a potential or quiescent inconsistency that can escalate into a real safety violation. Temporal inconsistency or the safety violation arises when we have the following:

$$L_6: \ Overbooked(f, d, \mathbf{t_2}, (\mathbf{t_4} - \mathbf{t_2} + k)), \text{ for some } k > 0.$$

Resorting to Algorithm 1 ($\mathbf{TID}_{LL}(L_5, L_6, [\mathbf{t_4}, (\mathbf{t_5} - \mathbf{t_4})], [\mathbf{t_2}, (\mathbf{t_4} - \mathbf{t_2} + k)]))$, we obtain the results below:

> temporal_relation: $OverlappedBy(\Delta_{L6}, \Delta_{L5})$;
> determination: $\nvDash_t(L_6, L_5)$;
> conflict_interval: $[\mathbf{t_4}, \mathbf{t_4} + k]$.

The interval relationship of $OverlappedBy(\Delta_{Overbooked}, \Delta_{\neg Overbooked})$ now results in an *overlapping* temporal inconsistency ($\nvDash_t(L_6, L_5)^o_{[t4, \ t4 + k]}$). When this is detected, it kicks in the aforementioned courses of actions by the airline that will resolve the safety violation before the flight takes off.

6 Related Work

There is in general a lack of reported work in the literature on characterizing temporal properties of knowledge base inconsistency. This, however, does not diminish the importance of the issue at stake. Of the published results, the work reported in [5] deals with the use of a paraconsistent temporal logic called QCTL for verifying temporal properties of inconsistent concurrent systems. In QCTL, temporal operators (*next, eventually, always,* and *until*) and path quantifiers (there exists a path, for all paths) are utilized to represent and reason about inconsistent specifications in concurrent systems. A model checking technique was proposed to verify the temporal properties of inconsistent concurrent systems over QCTL. There was no discussion on the types of inconsistency the approach can deal with.

7 Conclusion

In this paper, we delineate the concept of temporal inconsistency in knowledge that recognizes the fact that there are two important underpinnings for the real world inconsistent knowledge: logically conflicting and temporally coinciding. We also describe a novel approach in characterizing the temporal properties of KB inconsistency and in automatically detecting coinciding time intervals for temporally inconsistent propositions in a KB. The results also establish the semantic difference between the classical and temporal inconsistency. Relying on the interval temporal logic that underpins the reported work, our contribution lies in the fact that the results presented in this paper fill a gap in the temporal characterization of KB inconsistency.

Future work includes the following: empirical study on the proposed approach with regard to some deployed knowledge bases in applications (e.g., semantic web applications); tool suite to automate the detection process.

Acknowledgements. We would like to express our sincere appreciation to the editors of this special issue, and to the anonymous reviewers for their invaluable comments which greatly help improve the technical contents and the presentation of this paper.

References

1. Allen, J.F.: Toward a General Theory of Action and Time. Artificial Intelligence 23(2), 123–154 (1984)
2. Allen, J.F., Ferguson, G.: Actions and Events in Interval Temporal Logic. Journal of Logic and Computation 4(5), 531–579 (1994)
3. Bennett, B., Galton, A.: A Unifying Semantics for Time and Events. Artificial Intelligence 153(1-2), 13–48 (2004)
4. Brachman, R.J., Levesque, H.J.: Knowledge Representation and Reasoning. Morgan Kaufmann Publishers, San Francisco (2004)
5. Chen, D., Wu, J.: Model Checking temporal Aspects of Inconsistent Concurrent Systems based on Paraconsistent Logic. Electronic Notes in Theoretical Computer Science 157, 23–38 (2006)
6. Emerson, E.A.: Temporal and Modal Logic. In: van Leeuwen, J. (ed.) Handbook of Theoretical Computer Science. North-Holland Pub. Co., Amsterdam (1995)
7. Fagin, R., Halpern, J.Y., Moses, Y., Vardi, M.Y.: Reasoning about Knowledge. MIT Press, Cambridge (1995)
8. Gabby, D., Hunter, A.: Making Inconsistency Respectable 2: Meta-level Handling of Inconsistent Data. In: Moral, S., Kruse, R., Clarke, E. (eds.) ECSQARU 1993. LNCS, vol. 747, pp. 129–136. Springer, Heidelberg (1993)
9. Genesereth, M.R., Nilsson, N.J.: Logical Foundations of Artificial Intelligence. Morgan Kaufmann Publishers, Inc., Los Altos (1987)
10. Grant, J., Hunter, A.: Measuring Inconsistency in Knowledgebases. Journal of Intelligent Information Systems 27, 159–184 (2006)
11. Huang, Z., van Harmelen, F., ten Teije, A.: Reasoning with Inconsistent Ontologies. In: The Proceedings of the Nineteenth International Joint Conference on Artificial Intelligence, Edinburgh, Scotland, pp. 454–459 (2005)
12. Hunter, A., Nuseibeh, B.: Managing Inconsistent Specifications: Reasoning, Analysis, and Action. ACM Transactions on Software Engineering and Methodology 7(4), 335–367 (1998)
13. Hunter, A.: Measuring Inconsistency in Knowledge via Quasi-classical Models. In: The Proceedings of the National Conference on Artificial Intelligence, pp. 68–73 (2002)
14. Hunter, A.: Evaluating Significance of Inconsistencies. In: The Proceedings of the Eighteenth International Joint Conference on Artificial Intelligence, Acapulco, Mexico, pp. 468–473 (2003)
15. Hunter, A., Konieczny, S.: Approaches to Measuring Inconsistent Information. In: Bertossi, L., Hunter, A., Schaub, T. (eds.) Inconsistency Tolerance. LNCS, vol. 3300, pp. 189–234. Springer, Heidelberg (2004)
16. Knight, K.: Measuring inconsistency. Journal of Philosophical Logic 31(1), 77–98 (2002)
17. Knight, K.: A Theory of Inconsistency, Ph.D. Dissertation, Department of Mathematics, the University of Manchester, UK (2002)
18. Konieczny, S., Lang, J., Marquis, P.: Quantifying information and contradiction in propositional logic through test actions. In: The Proceedings of the Eighteenth International Joint Conference on Artificial Intelligence, Acapulco, Mexico, pp. 106–111 (2003)
19. Lee, D., et al.: Are Your Citations Clean? Communications of the ACM 50(12), 33–38 (2007)
20. Lenat, D.: The Dimensions of Context-Space, CYCorp Report (October 1998)
21. Leveson, N., Turner, C.S.: An Investigation of the Therac-25 Accidents. IEEE Computer 26(7), 18–41 (1993)

22. Levesque, H.J., Lakemeyer, G.: The Logic of Knowledge Bases. MIT Press, Cambridge (2000)
23. Manna, Z., Pnueli, A.: The Temporal Logic of Reactive and Concurrent Systems: Specification. Springer, New York (1992)
24. Menzies, T., Pecheur, C.: Verification and validation and artificial intelligence. In: Zelkowitz, M. (ed.) Advances in Computers, vol. 65, pp. 154–203. Elsevier, Amsterdam (2005)
25. Metric mishap caused loss of NASA orbiter,
    http://www.cnn.com/TECH/space/9909/30/mars.metric.02/
26. Rodriguez, A.: Inconsistency Issues in Spatial Databases. In: Bertossi, L., Hunter, A., Schaub, T. (eds.) Inconsistency Tolerance. LNCS, vol. 3300, pp. 237–269. Springer, Heidelberg (2005)
27. Rushby, J.: Quality Measures and Assurance for AI Software. NASA Contractor Report 4187 (October 1988)
28. Rushby, J., Whitehurst, R.A.: Formal Verification of AI Software. NASA Contractor Report 181827 (February 1989)
29. Sago Mine Explosion,
    http://www.cnn.com/2006/US/01/03/mine.explosion/index.html
30. Yin, X., Han, J., Yu, P.S.: Truth Discovery with Multiple Conflicting Information Providers on the Web. IEEE Transactions on Knowledge and Data Engineering 20(6), 796–808 (2008)
31. Zhang, D., Nguyen, D.: PREPARE: A Tool for Knowledge Base Verification. IEEE Transactions on Knowledge and Data Engineering 6(6), 983–989 (1994)
32. Zhang, D., Luqi: Approximate Declarative Semantics for Rule Base Anomalies. Knowledge-Based Systems 12(7), 341–353 (1999)
33. Zhang, D.: Fixpoint Semantics for Rule Base Anomalies. International Journal of Cognitive Informatics and Natural Intelligence 1(4), 14–25 (2007)
34. Zhang, D.: On Classifying Inconsistency in Autonomic Agent Systems, Technical Report, December 2007. Department of Computer Science, California State University, Sacramento (submitted for publication) (2007)

Images as Symbols:
An Associative Neurotransmitter-Field Model
of the Brodmann Areas

Douglas S. Greer

General Manifolds
Carlsbad, California, USA
dsgreer@gmanif.com

Abstract. The ability to associate images is the basis for learning relationships involving vision, hearing, tactile sensation, and kinetic motion. A new architecture is described that has only local, recurrent connections, but can directly form global image associations. This architecture has many similarities to the structure of the cerebral cortex, including the division into Brodmann areas, the distinct internal and external lamina, and the pattern of neuron interconnection. The images are represented as neurotransmitter fields, which differ from neural fields in the underlying principle that the state variables are not the neuron action potentials, but the chemical concentration of neurotransmitters in the extracellular space. The neurotransmitter cloud hypothesis, which asserts that functions of space, time and frequency, are encoded by the density of identifiable molecules, allows the abstract mathematical power of cellular processing to be extended by incorporating a new chemical model of computation. This makes it possible for a small number of neurons, even a single neuron, to establish an association between arbitrary images. A single layer of neurons, in effect, performs the computation of a two-layer neural network.

Analogous to the bits in an SR flip-flop, two arbitrary images can hold each other in place in an association processor and thereby form a short-term image memory. Just as the reciprocal voltage levels in a flip-flop can produce a dynamical system with two stable states, reciprocal-image pairs can generate stable attractors thereby allowing the images to serve as symbols. Spherically symmetric wavelets, identical to those found in the receptive fields of the retina, enable efficient image computations. Noise reduction in the continuous wavelet transform representations is possible using an orthogonal projection based on the reproducing kernel. Experimental results demonstrating stable reciprocal-image attractors are presented.

Keywords: Natural intelligence, computational neuroscience, cognitive signal processing, memory, pattern recognition, neurotransmitter fields.

1 Introduction

Throughout its entire extent, the cerebral cortex is composed of the same six cellular layers. Based on the relative thickness of these layers, Brodmann divided the cerebral

M.L. Gavrilova et al. (Eds.): Trans. on Comput. Sci. V, LNCS 5540, pp. 38–68, 2009.

cortex of humans into approximately fifty areas [1]. His illustration, shown in Fig. 1, was originally published one hundred years ago, but is still used in neuroscience textbooks today [2], [3]. Between Brodmann areas, the cortical layers vary in thickness, but within each area, they are roughly the same. At the boundaries between areas, there is a sharp and sudden change in the relative prominence of the six cellular lamina.

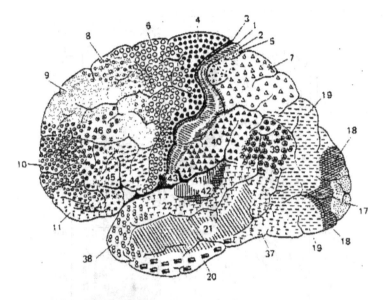

Fig. 1. Lateral view of the human brain divided into areas from the original work by Brodmann. The boundaries between the Brodmann areas are determined by the thickness of the six cellular layers.

A cross section of the neocortex, first published by Brodmann and Vogt, is shown as the background illustration in Fig. 2. The six cellular layers are labeled I through VI. Layers II and IV are the *external and internal granule* layers respectively, while layers III and V are the *external and internal pyramidal* layers. In addition, on the right side of Fig. 2, which has been prepared with a Weigert stain, the *external and internal Bands of Baillarger*, labeled 4 and 5b are visible. A complete theory of natural intelligence must somehow account for this uniform structure, and the existence of the two nearly identical (external and internal) parallel systems.

Also shown in the center of Fig. 2 is a loop representing a cortical column created from the local connections between neurons in the various layers. Even though most neurons have a width of only a few microns, they form layers of constant thickness over the full extent of the Brodmann areas, which may be several square centimeters in size. These facts are consistent with an image association model.

We use the term *image* or *field* in a general mathematical sense for a scalar or vector valued function defined on a manifold [4], [5], for example an open region of the plane $(\mathbb{R}^2)$. We use script letters to denote function transforms, for example the wavelet transform of a function f is written as $\mathcal{W}[f]$.

Each layer of each Brodmann area can be considered to be one image. Consequently, the general notion of an image plays a role in cognition that extends beyond computer vision. For example, we can model an audio spectrogram that represents time in the horizontal dimension and frequency in the vertical dimension as a two-dimensional image. Similar tonotopic maps exist at several locations in the central nervous system (CNS) [2], [6].

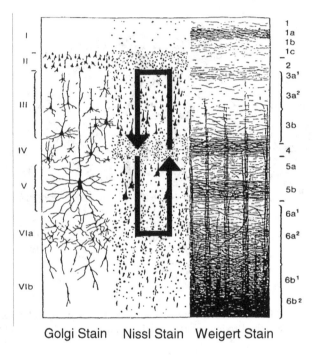

Golgi Stain   Nissl Stain   Weigert Stain

Fig. 2. A cross-section of the cerebral cortex showing the six cellular layers. Most of the structures occur in external/internal pairs such as the granular layers (II and IV), the pyramidal layers (III and V) and the Bands of Baillarger (4 and 5b). The loop in the center depicts the connections of a single cortical column. (background from Brodmann & Vogt, 1908).

The generalization of images to include any function defined on a manifold is a powerful one in terms of its ability to describe and model many natural phenomena. Images can include mass (external world), position (location of the body surface), energy (visible light and sound vibrations), and force (pressure on the body surface, or the tension on the combined muscle cross-sections).

For example, the surface of the skin (a somatotopic map) can be modeled as a two-dimensional manifold, and its position in space then becomes a vector valued function defined on that surface. A manifest of the two-dimensional images that are processed by the CNS is extensive [7], and includes retinotopic maps, spatial maps, and kinetic maps.

In this context, learning and retrieving image associations can be viewed as a basic cognitive function. As an example, when children learn a new word, say "horse", they learn to associate two images: the visual image of a picture of a horse and the two-dimensional audio spectrogram of the sound of the word "horse".

An assumption of the standard neural-network perceptron model is that the electrical action potentials are the information, and neurotransmitters serve only as a mechanism for transferring a weighted response between neurons. However, if we step back and allow ourselves to view the neurotransmitter concentration as the data, a different model of computation emerges.

Information and computation are mathematical abstractions and can be realized by many different physical means. For example, mechanical computing devices have a long history that predates the discovery of electricity. In complex biological organisms, cellular signaling generally takes place using chemical mechanisms. If a limb of a tree is removed, the tree will grow a new one with the all the detailed structure necessary to create the leaves and branches. This massive computation takes place without electricity. The chemical signals employed in this computation are virtually identical to the substances used as neurotransmitters in the brain.

To survive, an organism must maintain efficient and accurate representations of force, mass, energy, and other functions of time, space, and frequency. We can imagine a compact, high-resolution data representation analogous to ink on a sheet of paper. The neurotransmitter "ink" is written and erased by the axons and "read" by the dendrites. We explore this new model of natural computation where the data itself is chemical rather than electrical. The information is encoded by the molecular density of neurotransmitter that is released by multiple axons into the synaptic cleft.

Modeling a large number of particles with integral and differential equations is common in physics. Neurotransmitter fields use this same abstraction to model data as a continuous function in three-dimensional space. The need for a continuous rather than discrete cognitive representation also follows from the scientific philosophy described by Monad [8]. As applied to neuroscience, any inconsistencies that may appear due to the discretization of physical laws will tend to be removed by natural selection. Therefore, since space is isotropic and homogeneous, at some level of abstraction the computational model itself should be continuous. Since these calculations may be carried out by many types of biological cells other than neurons, we use the term *computational manifold* to refer to the generalization of discrete neural networks to continuous spaces.

We discuss two types of image association processors. The first of these, the Λ-map, is feed-forward and has no recurrent loops. It can however perform logic operations by producing the output image that is associated with multiple input images. The second type of association processor, the *psymap*, is constructed from two Λ-maps and is analogous to a set-reset flip-flop where the individual bits have been replaced by two-dimensional images. Its design uses recurrence to integrate the results of many locally connected processing elements into an overall global image association.

Borrowing from the terminology of digital design, specifically a programmable logic array, the term *psymap array* is used to specify an interconnected collection of psymaps where the bits are again replaced by images.

The coefficients of spectral representations such as wavelets, in effect, encode weighted averages over regions. Spherically symmetric wavelets, similar to the receptive fields of the retina, allow image associations to be computed faster and more efficiently. We hypothesize that these wavelet-like fields are not unique to the visual system but are a general principle of operation that governs the entire cerebral cortex.

The continuous wavelet transform contains redundant information that is characterized by the reproducing kernel [9]. When associations are formed between image spectral representations, this redundant information can be used to reduce errors and increase stability.

The behavior of a psymap over time can be viewed as a dynamical system. We describe an algorithm where two arbitrary images can be linked together to form a *reciprocal image* pair. Each pair can be used to create a *reciprocal-image attractor* where images near an attractor converge toward it.

Symbolic processing is traditionally performed by manipulating bit patterns in a digital computer. We present a new type of symbolic processing where the symbols are images. Whether the symbols are discrete bit patterns or continuous functions, in order to overcome the effects of noise, a necessary condition required to construct actual implementations is that the symbol representations act as stable fixed-points in a dynamical system.

A new software package was written to verify that neurotransmitter-field theory could be used to construct structurally stable reciprocal-image attractors. The computational results shown demonstrate how stable attractor basins can be created around arbitrary image pairs and thereby form the foundation of an image-based symbol processing system.

1.1 Related Work

A psymap array is an implementation of an abstract symbol processing system. Consequently, descriptions of cognitive models that can be expressed in terms of symbolic operations [10], [11] can be used to evaluate psymaps, which use images to represent symbols. Control masks, in particular time-varying control masks, act as instructions for directing the operation of the array. Therefore, the ability to recall control mask images translates into the ability to recall procedures that operate on symbols. Cognitive informatics describes how the acquisition, representation, retrieval, and comprehension of concepts may be expressed as an aggregation of symbols and operators [12], [13]. These theoretical frameworks can be applied in the analysis of psymap arrays as computational engines that evaluate semantic functions.

In contrast to some previous work that uses feature detection and extraction and performs the analysis in a lower dimensional space, we discuss algorithms where the images maintain their topological structure and the associations take place directly between the image wavelet transforms.

Willshaw [14] first proposed a model of associative memories using non-holographic methods. Bidirectional associative memories were originally proposed by Kosko [15] and subsequently developed by others [16]. These neural networks are designed such that all of the outputs directly depend on all the inputs, that is, they require global connections to the entire input image. Consequently, implementations can suffer from poor performance for high-resolution images or large training sets. Cellular neural networks, first described by Chua and Yang [17], are implemented with local connections, but do not use spectral methods or aggregate multiple images with a unified control structure.

A second type of neural-network associative memory is created by following the trajectory of a particle in a dynamical system governed by a recurrence of the form $\vec{x}_{i+1} = f(\vec{x}_i)$. The association is not fixed in time but is the relationship between a starting point and a final fixed point of the dynamical system. To distinguish between these two types of association mechanisms we will refer to the second type as a classification, where each unique fixed point identifies a *class* that equals the set of all points whose trajectory leads to that point. The psymap model may also go through a classification phase, but once the association is formed, the relationship between the associated images is static and explicit.

Wavelet networks combine the functional decomposition of wavelets with the parameter estimation and learning capacity of neural networks [18]–[20]. However, these differ from the psymap methods in their lack of reciprocal associations formed between representations. Moreover, their use of wavelet basis functions rather than the continuous wavelet transform implies that the projection based on the reproducing kernel will not reduce noise and add stability since the basis functions are by definition linearly independent.

2 Neurotransmitter Fields

Neural fields have been studied for a long time [21]–[24] and comprehensively reviewed by several authors [25], [26]. This approach models the behavior of a large number of neurons by taking the continuum limit of discrete neural networks where the continuous state variables are a function in space representing the mean firing rates.

The distinction between neural fields and neurotransmitter fields is the physical quantity under consideration. Neural fields attempt to model the spatial distribution of mean neuron-firing rates as a real-valued function, while neurotransmitter fields model the concentration of neurotransmitters in the extracellular space as a real-valued function. The localization of neurotransmitters to the space within the synaptic cleft is seen as an evolutionary adaptation that limits diffusion and increases the efficiency.

In order to develop the theory, we put forth a single proposition: the neurotransmitter cloud hypothesis. Empirical evidence and deductive arguments are provided which support this proposition, but verification will require further investigation and analysis. Acceptance of the hypotheses, like including an additional mathematical axiom, allows us to explore a new computational model that characterizes the electro-chemical properties of the neuron.

2.1 Evolution of the Nervous System

Although the evolution of the senses and the central nervous system was a complex process that occurred over an extended time interval [27], we can attempt to understand some of the general constraints that may have influenced its development. One of these constraints was the need to evaluate the current state of the body and its immediate environment. This required the creation of internal representations that could be equated with physical quantities defined over the continuous variables of space, time, and frequency.

In the standard neural network model, a synapse is characterized mathematically by a single real-valued weight representing the effect one neuron has on another. The products of the weights times the activation values of the input neurons are summed, and a nonlinear transfer function is applied to the result [28]. This model describes electrical and chemical synapses uniformly, that is, by a single real value. Examining the difference between electrical and chemical synapses, we note that electrical synapses, which may have a weighted response proportional to the number of ion channels connecting the pre- and postsynaptic neuron, are more than ten times faster. They are also more efficient, since they do not require the metabolism of neurotransmitters, or the mechanics of chemical signaling. However, chemical synapses are found almost exclusively throughout the central nervous systems of vertebrates. This raises the question: Given a time interval of several hundred million years, and the wide range of species involved, why has nature consistently retained the cumbersome chemical synapses and not replaced them with electrical synapses?

We note that neurotransmitters, the core component of chemical synapses, are actually located outside the neuron cell walls in the extracellular space. Moreover, the chemical signaling often occurs in multiple-synapse boutons such as the one shown in Fig. 3. Within these complex synapses, which connect the axons and dendritic spines of many adjacent neurons, the density of neurotransmitter is equal to the sum of the contributions from each of the individual axons.

Another constraint during the course of evolution was the limited amount of processing power available. Solutions that required more than a very small number of neurons were not feasible. In addition, the space within the organism that could be devoted to representing physical quantities was limited, so small, compact representations were preferable.

If we leave the confines of the standard neuron model and consider the density of neurotransmitters as the state variables, we discover a number of advantages. The first is higher resolution; billions of small molecules can fit in the space occupied by a single neuron. The second is energy consumption; the concentration of neurotransmitters, like the concentration of ink on a sheet of paper, is passive and can store information indefinitely without expending energy. In contrast, action potentials require the continuous expenditure of energy in order to maintain state. Another advantage is that a very high-resolution representation can be maintained with only a few processing elements. For example, the terminal arbor of a single neuron that encodes a joint angle can support a high-resolution model describing the location of the surface of the limb in space. This collection of related concepts results in the following conjecture.

The Neurotransmitter Cloud Hypothesis: When multicellular fauna first appeared, organisms began to represent quantities such as mass, force, energy and position by the chemical concentration of identifiable molecules in the extracellular space. The basic principles of operation developed during this period still govern the central nervous system today.

The basic laws of physics are based on quantities defined in space, time, and frequency, which can be internally represented by the chemical concentration of neurotransmitters in three-dimensional space.

Neurotransmitter clouds in early metazoa would have suffered from two problems: chemical diffusion of the molecules and chemical inertia due to the large amount of neurotransmitter required to fill in the extracellular space. As a result, evolutionary adaptation would have favored neural structures where the neurotransmitters remained confined to the small regions in the synaptic clefts between the pre- and postsynaptic neurons.

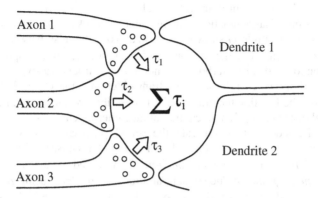

Fig. 3. This idealized view of multi-synapse boutons shows how the concentration of neuro-transmitter perceived by multiple dendrites is the summation of that produced by three separate axon terminals. The summation occurs in the extracellular space and is separated from the intracellular summation by the nonlinear responses of the cell membranes.

In order to visualize how a computation can be performed on a neurotransmitter cloud, imagine the dendritic arbor of a neuron as a leafless tree with its branches inside of the cloud. The surface of the tree is "painted" with a shade of gray that corresponds to its sensitivity to a particular neurotransmitter. When multiplied by the actual concentration of neurotransmitter present in the extracellular space, and integrated over a region of space that contains the dendritic tree, the result is a first-order approximation of the neuron's response. We can mathematically represent the "shade of gray" that corresponds to the sensitivity of a neuron's dendritic arbor in physical space by a function $\mu(x,y,z)$.

2.2 Neurotransmitter Field Theory

The standard projection neural network calculation is based on the inner product of two vectors, a vector of input values, and a weight vector. Hilbert spaces generalize the inner product operation to continuous domains by replacing the summation of the products of the vector coefficients, with the integral of the product of two functions [29]. One of these two functions, $\mu(x,y,z)$, is used to represent the sensitivity of the dendritic arbor and is analogous to the weight vector.

Let H be the three-dimensional space representing a neurotransmitter cloud, and let $h(x,y,z)$ be a field corresponding to the density of transmitter in the extracellular space. We conceptually model the operation of a neuron as an abstract *Processing Element* (PE). The dendritic arbor computation of the PE, which is analogous to the vector inner product, is defined by the integral of h and with respect to μ

$$\text{response} = \iiint_H h(x, y, z)\, d\mu(x, y, z) \tag{1}$$

In [30] it is shown that by using Lebesgue integrals and Dirac delta functions the mathematical formulation of neurotransmitter fields (1) subsumes the functionality of the neural networks. That is, for every neural network, there exists a corresponding neurotransmitter field integral over a dendritic tree that generates an identical result.

Mathematically the neurotransmitter "clouds" are three-dimensional manifolds which we illustrate diagrammatically as rectangular blocks such as the input manifold H and the output manifold G shown in Fig. 4A. To distinguish between the input and output spaces, we substitute the parameters (ξ,η,ζ) for (x,y,z) in the input manifold H.

In addition to the dendritic arbor, each neuron (or astrocyte) also has an axonal tree, or terminal arbor, which releases neurotransmitter into the extracellular space. Let $\tau(x,y,z)$ denote the function that quantitatively describes the output of a neuron in terms of the spatial distribution of chemical neurotransmitter it generates.

We use the index i to enumerate the set of processing elements $\{PE_i\}$. Each processing element, such as the ones shown in Fig. 4A, consists of a unique *receptor measure*, $\mu_i(\xi,\eta,\zeta)$, a *transmitter function* $\tau_i(x,y,z)$ and nonlinear *cell-membrane-transfer functions*, χ and σ. The spatial variations are modeled using μ_i and τ_i and we assume that χ and σ are fixed functions of a single real variable, similar to the sigmoidal transfer functions used in neural networks.

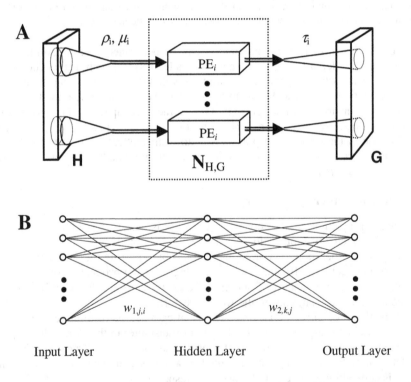

Fig. 4. A computational manifold (A) transforms continuous fields while a neural network (B) transforms discrete vectors. In the neurotransmitter field model (A) the Processing Elements (PE_i) represent neurons which are points in the function space $N_{H,G}$. Since there are effectively two summations, one in the intracellular space and one in the extracellular space, the receptor measure μ, together with the transmitter function τ, allow a single layer of neurons to perform the equivalent computation of a two-layer neural network The nonlinear responses of the cell membranes separate the two summations.

The transformation from a discrete real value back to a continuous field results from scaling the output of the nonlinear transfer function σ by the transmitter function $\tau_i(x, y, z)$. Taking into account the cell-membrane transfer functions and summing over all of the PEs gives the complete output function g.

$$g(x, y, z) = \chi \left(\sum_i \sigma \left(\iiint_H h(\xi, \eta, \zeta) \, d\mu_i(\xi, \eta, \zeta) \right) \cdot \tau_i(x, y, z) \right) \qquad (2)$$

The receptor measure μ_i models the shape and sensitivity the dendritic arbor in the input manifold H, while transmitter function τ_i models the signal distribution in the terminal arbor and the concomitant release of neurotransmitters into the output manifold G. The receptor measures $\{\mu_i\}$ and the transmitter functions $\{\tau_i\}$ perform the complementary operations of converting back and forth between fields defined on continuous manifolds and discrete real values resulting in both an integral and a summation sign in (2).

2.3 Basis Functions

The continuous version of a projection neural network can be extended by generalizing the notion of radial basis functions [31] to computational manifolds. For discrete neural networks, a set of pattern vectors $\{u_\alpha\}$ and a radial basis function θ form the discriminate functions $\theta(\|u - u_\alpha\|)$. The real-valued function $\theta(x)$ has its maximum at the origin and the properties $\theta(x) > 0$ and $\theta(x) \to 0$ as $|x| \to \infty$. Typically the theta functions are a Gaussian, $\exp(-x^2/2(stddev)^2)$, or a similar function.

To construct the analogous continuous basis functions, we replace the discrete pattern vectors u_α with a continuous field ρ_α. Each of the functions $\rho_\alpha(\xi, \eta, \zeta)$ represents a "pattern" density defined on the input manifold. We associate a particular "target" function $g_\alpha(x,y,z)$ in the output manifold with each input pattern ρ_α. Assuming that there are several PEs available for every pattern, each PE_i is assigned a particular pattern which we label ρ_i.

The equation corresponding to a basis-function neural network can be obtained by substituting $\theta(h(\xi, \eta, \zeta) - \rho_i(\xi, \eta, \zeta))$ for h in (2)

$$g(x, y, z) = \sum_i \sigma \left(\int_H \theta(h - \rho_i) \, d\mu_i \right) \cdot \tau_i(x, y, z) \qquad (3)$$

where we have omitted the variables of integration (ξ, η, ζ) for h, ρ_i, and μ_i.

Each processing element now has an additional property ρ_i, which represents the pattern to which it is the most sensitive. For each PE, the integral inside (3) is maximum when $h = \rho_i$ over the region of integration. This in turn maximizes the coefficient for the transmitter function τ_i. The sum of the transmitter functions $\{\tau_i\}$ associated with a particular input pattern ρ_α can then be defined to approximate the desired target function g_α, thereby creating the required associations.

The measures μ_i in (2) and (3) can identify the regions where the pattern ρ_i is the most sensitive. For example, we can imagine photographs of two different animals that appear very similar except for a few key features. The photographs, representing two patterns ρ_1 and ρ_2, are approximately equal, but the measures can be trained so that their value where the patterns are the same is small, but in the key regions where the patterns differ, they have much larger values. In this way, even though the two image patterns are almost the same, the output functions g_α that result from the integrals in Equation (3) could be very different.

2.4 Computational Equivalence

While models that use action potentials as state variables can form associations by using matrix operations on a large vector of neuron outputs, equation (3) shows the neurotransmitter-field state model makes it possible for a small number of neurons, even a single neuron, to establish an association between an arbitrary input pattern $\rho_\alpha(\xi,\eta,\zeta)$ and an arbitrary output pattern $g_\alpha(x,y,z)$.

A continuous computational manifold and a two-layer discrete neural network are shown in Fig. 4A and 4B. The measures $\{\mu_i\}$ in the computational manifolds can replace the weights $\{w_k\}$ in the neural network. The corresponding summation takes place inside the cell. Since the transmitter functions $\{\tau_i\}$ can extend over a large area, even the entire output manifold, many different processing elements may contribute to the concentration of neurotransmitter at any particular point (x,y,z). Consequently, the summations in (2) and (3) are *also* equivalent to the summations in a neural network where the weights correspond to the values of the transmitter functions at a given point. This summation takes place outside the cell, as illustrated in Fig. 3. Since both the integrals with respect to the measures μ_i, and the summations over the transmitter functions τ_i, perform operations analogous to the inner product with a weight vector in a single-layer neural network, together they perform an operation analogous to a two-layer neural network.

2.5 Function Spaces

The nodes of the neural network shown in Fig. 4B are partitioned into the input layer, the hidden layer, and the output layer. In the computational manifold model shown in 4A, the input layer is analogous to the input manifold H, and the output layer is analogous to the output manifold G, where both h and g define the continuous distribution of neurotransmitters in physical space. The "hidden" layer is the space $\mathbf{N}_{H,G}$, which equals the Cartesian product of two function spaces: the space of all possible (receptor) measures on H, and the space all possible output (transmitter) functions on G. The individual neurons, PE_i, are points in this infinite-dimensional product space.

When samples of a continuous function defined on a high-dimensional space are arranged in a lower dimensional space, they will in general appear to be discontinuous. Consequently, when a collection of processing elements, $\{PE_i\}$, representing samples taken from the infinite-dimensional function space $\mathbf{N}_{H,Q}$ are arranged in three-dimensional physical space, the outputs will seem discontinuous. Moreover, realistic neural field models that attempt to describe the observed firing rates of large groups of neurons as a continuous function in physical space will be difficult or impossible to create. It may be expected that the responses of neighboring neurons will be relatively uncorrelated even when the underlying neurotransmitter fields are continuous.

2.6 Cellular Computations and Neuroglia

An abstract processing element can be defined as any cell that detects and emits a signal. In biological systems, we cannot restrict the focus to electrical signals, since even neurons detect chemical signals. Moreover, we cannot expect the behavior of neurons to be predicable if neighboring cells that use only chemical signaling are excluded from consideration.

In the central nervous system of vertebrates, there are 10 to 50 times more glial cells than neurons [2]. Astrocytes, the most common type of neuroglia, are receptive to potassium ions and take up neurotransmitters in synaptic zones. Glial cells have also been shown to release neurotransmitters.

Unlike neurons, glial cells do not generate action potentials. Consequently, if state is encoded in the firing of neurons, glia are relegated to a support role. However, in a neurotransmitter-field model, glia can take a central position along side neurons. They may participate in both short-term and long-term memory as well as computations. However, since they lack action potentials, glial cells transmit the results of their computations more slowly.

A notable difference between a perceptron neural-network model and the neurotransmitter-field model is the speed and simplicity of learning new associations. A two-layer neural network such as the one shown in Fig. 4B typically requires an extended training phase involving repeated iterations of a back-propagation algorithm. In contrast, the neurotransmitter field model shown in Fig. 4A can learn a new association by adding a single cell, which could be a glial cell. When the new cell recognizes its image pattern, ρ_i, in the input manifold, it emits the associated neurotransmitter τ_i into the output manifold. If the new input pattern is relatively unique, previously learned associations will not be affected.

3 Association Processors

Based on neurotransmitter-field theory, we can construct a system where arbitrary collections of photographs operate as symbols. If we wish to represent numbers or letters, we can pick a font and record images of their alphanumeric glyph. Similarly, fixed-time snapshots of any two-dimensional computational map can be used as symbols representing external stimuli or motor control actions.

In Fig. 5, we show how using topological alignment, multiple input images can be combined and associated with a single output image. The ability to associate images allows us to learn the addition tables, multiplication tables or arbitrary Boolean logic operations. We therefore refer to this feed-forward system as a Λ-*map* (Λ from the Greek word *Logikos*). Like a Boolean logic gate, this computational model is stateless; its output depends only on the current value of its inputs.

The performance of image association processors quickly begins to degrade for high-resolution images when every pixel in the input image is directly used in the calculation of every output pixel. We therefore create a system with a *lattice* of processing elements where each PE takes its inputs from only a local area.

We define the *support* of PE_i to be those points in the image where the measure μ_i is non-zero. If the support of a PE is contained in a region of radius d, where d is small relative to the size of the image, we say the PE has *local support*. Although neighboring PEs will in general have overlapping regions of support, the design does not restrict the size or shape of the regions. Fig. 5 shows a PE where inputs are taken from three separate images, but each image has a different sized region of support, with one relatively "narrow", and another relatively "broad".

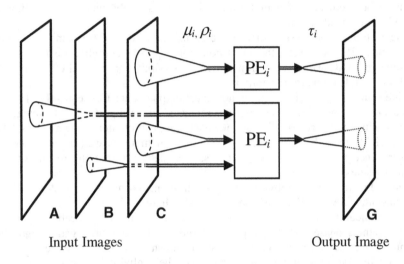

Input Images   Output Image

Fig. 5. The Λ-map produces an output image that is associated with multiple input images. While the input and output images are neurotransmitter fields, the model is stateless in that its current output does not depend on the previous inputs. Processing elements with local support take their inputs from small regions in multiple input images. The regions of support may be narrow as shown for image *B* or broad as shown for image *C*. The output image is produced by a lattice of PEs operating in parallel.

As shown, the output image will in general consist of a melee of uncoordinated results, with PEs in different areas recognizing different patterns. However, using recurrence the system can form numerous cyclical loops which integrate the results of the individual processing elements into a coherent overall image association.

3.1 The Psymap

Character strings that serve as symbols in a digital computer are composed of a finite sequence of bits. A binary digit is an abstract mathematical concept, which in practice, is implemented as bi-stable analog circuit. The basic unit of computer memory, the Set-Reset (SR) flip-flop shown in Fig. 6A, has two stable states corresponding to "0" and "1". Initial voltage values move along a path toward one of the two *attractors*. The set of points whose paths lead to an attractor is referred to as its *attractor basin*. Since the system must operate correctly in the presence of noise, the attractor basin must contain an open neighborhood around each *stable* fixed point. If the mathematical topology does not change under small variations of the defining parameters, the dynamical system is considered to be *structurally stable* [32]. Thus, in the physics of computer circuits, a symbol is actually a stable fixed-point-attractor basin of a dynamical system.

Internally, inside each bit of static random access memory (SRAM) is a logic circuit equivalent to the SR flip-flop shown in Figure 6A. If the S or R inputs, which are normally one, momentarily go to zero the flip-flop will set or reset and remain there until either the S or R input is changed again. We replace the single bits S, R, Q and Q' in the SR flip-flop with two-dimensional images and we replace the two NAND gates with two Λ-maps. The resulting *psymap* is shown in Fig. 6B, where the output Q' has been relabeled as P.

A   **B**

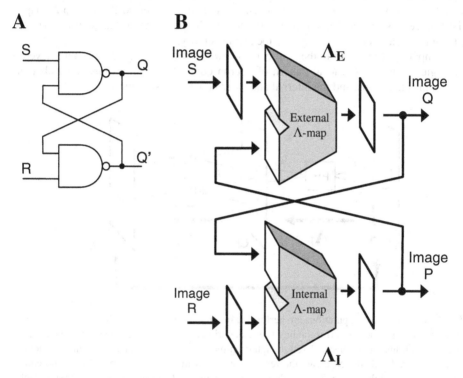

Fig. 6. The *Psymap* (*B*) is analogous to the SR flip-flop (*A*) where the bits have been replaced by images and the NAND gates have been replaced by Λ-maps. The two Λ-MAPs are labeled *External* and *Internal* and correspond to the external in internal lamina of the cerebral cortex.

Referring to the psymap in Fig. 6B, let $\Lambda_E(S, P) = Q$ denote the external Λ-map, and let $\Lambda_I(R, Q) = P$ denote the internal Λ-map. Let *Null* denote a predefined "blank" image, and let $\{(a_1, b_1), (a_2, b_2) \dots (a_i, b_i) \dots (a_n, b_n)\}$ be an arbitrary collection of n image pairs. Suppose we program Λ_E such that $\Lambda_E(Null, b_i) = a_i$ for all i, and program Λ_I such that $\Lambda_I(Null, a_i) = b_i$ for all i. When the R and S inputs are Null, the psymap will have n fixed-points corresponding to the n image pairs (a_i, b_i). We will refer to the images that form an image pair (a_i, b_i) as *reciprocal* images.

In addition to the above, suppose we have n input images $(s_1, s_2, \dots , s_n)$ and we add the additional associations to Λ_E such that $\Lambda_E(s_i, X) = a_i$ for any image X. Then by changing the S input from Null to s_i, we can force the psymap from whatever state it is currently in to the state identified by the reciprocal image pair (a_i, b_i). If the fixed-point is stable, when the S input returns to Null, the psymap will remain in this state, until either the S or R input image changes again.

In a simple feed-forward image association system with local support there is no way to coordinate the individual PE outputs to form a consistent global image association. However, the two Λ-maps which make up psymap form an image loop. The output of any particular PE feeds into the local support of several other PEs in the opposite Λ-map. These in turn form many local loops by feeding back into the original PE. Figure 7 illustrates how this can occur by showing connected PEs in two Λ-maps. This process is repeated in an interconnected mesh and is the critical step in forming global image associations using only local connectivity.

Comparing Fig. 7 with the cortical columns shown in Fig. 2, we note several analogies including the local pattern of neuron connections, the extensive number of interconnected loops, and the internal and external laminar structure.

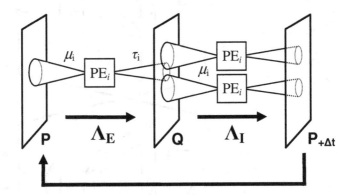

Fig. 7. The psymap computations are performed by processing elements with overlapping local input and output regions corresponding to the dendritic and axonal trees, where the image data is the neurotransmitter density. Even though each receptor measure μ_i, and transmitter function τ_i, cover only local limited areas, the recurrence allows their effect to spread over the entire image. The far right image $P_{+\Delta t}$ equals the leftmost image P following one processing cycle lasting time Δt. The pattern of interconnection closely resembles the cytoarchitecture of the cortical columns shown in Fig. 2.

In the section on experimental results, we demonstrate an implementation of the system shown in Fig. 7, which shows how local information in the input image can progressively spread to wider and wider areas as the recurrence progresses. This proves that processing elements with local support can be used to create a psymap with *stable global reciprocal-image attractors*. Consequently, since both of the associated images can be chosen arbitrarily, they can serve symbols in a cognitive system.

3.2 Receptive Fields

Recursion allows processing elements with local connections to form global associations. Varying the size of regions from which the neurons make connections, such as those shown in Fig. 5, allows a psymap to find the matching image more readily. We could "pre-compute" the average of small image regions with various sizes and use these averages as the inputs to the PEs. Rather than simply using averages, we can extend this idea by using local spectral functions such as two-dimensional wavelets.

We can associate images by associating their wavelet transforms. The spectral functions used could be identical to the receptive fields in the retina of the eye.

Axons from the *on* and *off-center ganglion cells* in the retina form the optic nerve. These cells receive inputs from nearby photosensitive rods or cones in a roughly circular *receptive field* that contains both a center area and a surrounding ring of opposite polarity. The receptive fields differ in size and are similar to spherically symmetric functions where the variation in the receptive field size is given by s, the *scaling* parameter [33]. The axons corresponding to the cells with large and small receptive fields remain segregated as they project to the lateral geniculate nucleus and on to the primary visual cortex (Brodmann area 17) [2], [34]. On coordinate system in the primary visual cortex, with the z-axis perpendicular to the surface, the size (scale) of the receptive field is topographically mapped to the z dimension. Throughout the cerebral cortex, the layers which make up the cortex have a multi-cellular thickness in the perpendicular z direction that gives them a three dimensional structure.

Spectral psymaps are based on the following hypothesis: The topographic mapping of receptive fields increasing in size onto the third (z) dimension is not an anomaly unique to visual processing, but rather, a general principle of operation that improves the efficiency and performance of the association computations throughout the entire cerebral cortex.

3.3 Wavelets

In one-dimension, a single main wavelet φ, which is normalized and symmetric about the origin, can generate a family of wavelets at position x and scale s.

$$\varphi_{x,s}(\xi) = \frac{1}{\sqrt{s}} \varphi\left(\frac{\xi - x}{s}\right) \tag{4}$$

Wavelets with a relatively large value of scaling parameter s act as low-frequency wavelets, analogous to broad areas of support, and those with a relatively small value of s act as high-frequency wavelets, analogous to narrow regions of support.

The wavelet transform of a function f is given by

$$\mathcal{W}[f](x,s) = \int_{-\infty}^{+\infty} f(\xi)\varphi_{x,s}^{*}(\xi)\,d\xi = \langle f, \varphi_{x,s}\rangle \tag{5}$$

where φ^{*} denotes the complex conjugate of φ.

Several methods are available for generating sets of multidimensional wavelets whose linear combinations are dense in $\mathbf{L}^2(\mathbb{R}^n)$. Of particular interest are spherically symmetric wavelets, which can be expressed in the form $\varphi(\mathbf{x}) = f(\|\mathbf{x}\|); \mathbf{x} \in \mathbb{R}^n$ for some one-dimensional function f. The scale parameter for spherically symmetric wavelets is a single real-valued positive number $s \in \mathbb{R}_+$. Therefore, the overall parameter space has dimension $n+1$. While some wavelets, such as the normalized second derivative of a Gaussian function, have non-zero values out to infinity, we are mainly interested in wavelets with compact support [35], in particular wavelets that are zero outside a small local region.

If a wavelet is spherically symmetric, so is its Fourier transform. Thus, $\hat{\varphi}(\omega) = \gamma(\|\omega\|)$ for some function γ, and the admissibility condition [9] is

$$C_\chi = \int_0^{+\infty} \frac{|\gamma(\omega)|^2}{\omega} d\omega < \infty. \tag{6}$$

For $f \in \mathbf{L}^2(\mathbb{R}^n)$ the wavelet transform, $\tilde{f}$, is defined by extending the integral in (5) to n dimensions. The inverse wavelet transform in n dimensions [36] is given by

$$f(\xi) = \mathcal{W}^{-1}[\tilde{f}](\xi) = \frac{1}{C_\chi} \int_{\mathbb{R}_+} \int_{\mathbb{R}^n} \tilde{f}(\mathbf{x}, s)\varphi_{\mathbf{x},s}(\xi)\, d\mathbf{x} \frac{ds}{s^{n+1}}. \tag{7}$$

Wavelet transforms on spherical surfaces are widely used in science and can be defined on other manifolds as well using the inner product [35].

3.4 Spectral Psymaps

Each of the Brodmann areas has a unique shape that projects along the perpendicular direction of the cortical columns onto a two-dimensional manifold $\mathbb{M}_i$. The spectral functions within each cortical layer are parameterized by a center position (x,y) and a single real-valued scale factor $s \in \mathbb{R}_+$. The resulting three-dimensional manifold $\mathbb{J}_i = (\mathbb{M}_i \times \mathbb{R}_+)$, maps to a single cortical layer of a single Brodmann area.

Examining the psymap shown in Fig. 7, we now replace the two-dimensional images with three-dimensional "slabs" of thickness z_0. The scaling parameter $s \in (0,\infty)$ is monotonically mapped to the interval $(0, z_0)$ that corresponds to the physical thickness of a single cortical layer in a particular Brodmann area. In the resulting three-dimensional spectral manifold, the input is now taken from a local volume where the values represent the coefficients of specific spectral functions. The PE dendritic tree spans small regions of the three-dimensional input spectral manifolds, and transmits results to a three-dimensional output spectral manifold.

Low-frequency spectral functions measure components over a large area of an image. Consequently, even though a processing element has only local connections near an image point (x_0, y_0), if the connections extend through the entire thickness of the cortical layer, its output value can change based on changes in the input image from the finest to the coarsest levels of detail. Moreover, the recursion in the psymap allows the output values of PEs that correspond to low-frequency spectral functions to propagate quickly throughout the entire psymap.

3.5 Orthogonal Projections

Because of the scale parameter s, the spectral manifold on which $\mathcal{W}[f](x, y, s)$ is defined is three-dimensional. Since this spectral manifold has a higher dimension than the original two-dimensional image space, there are many spectral functions for which there is no corresponding image. Mathematically, almost all functions h, defined on the space of continuous wavelet transforms, do not have a well-defined inverse $\mathcal{W}^{-1}[h]$.

Consequently, the space of transformed functions is over-specified and contains redundant information. This redundancy can be used to reduce errors and make the association process faster and more efficient.

The PEs compute the required outputs for a large number of stored associations based on only limited local information. Consequently, the overall result of these calculations can only be an approximation, which may not have a well-defined inverse. However, using the *reproducing kernel* it is possible to estimate the approximation error and calculate the closest function for which the inverse spectral transformation exists.

For the one-dimensional case, the following equation defines the necessary and sufficient conditions for a function $\mathcal{W}[f]$ to be a wavelet transform [9].

$$\mathcal{W}[f](x,s) = \frac{1}{C_\varphi} \int_0^{+\infty} \int_{-\infty}^{+\infty} \mathcal{W}[f](\xi,\eta) K(x,s,\xi,\eta) \, d\xi \, \frac{d\eta}{\eta^2} \tag{8}$$

where the constant C_φ is given by (6). The reproducing kernel K measures the correlation between the wavelets $\varphi_{x,s}(\alpha)$ and $\varphi_{\xi,\eta}(\alpha)$ and is defined by

$$K(x,s,\xi,\eta) = \int_{-\infty}^{+\infty} \varphi_{x,s}(\alpha) \varphi_{\xi,\eta}(\alpha) \, d\alpha = \langle \varphi_{x,s}, \varphi_{\xi,\eta} \rangle. \tag{9}$$

Let $H = \mathbf{L}^2(\mathbb{R} \times \mathbb{R}_+)$ and let U denote the linear subspace of H where the inverse wavelet transform exists. Using the reproducing kernel specified by (9) we define the linear operator $\mathcal{V}$ by

$$\mathcal{V}[f](x,s) = \frac{1}{C_\varphi} \int_0^{+\infty} \int_{-\infty}^{+\infty} f(\xi,\eta) K(x,s,\xi,\eta) \, d\xi \, \frac{d\eta}{\eta^2} \tag{10}$$

From (8) we note that for $f \in U$, $\mathcal{V}[f] = f$. In a straightforward proof, it can be shown that $\mathcal{V}$ is an orthogonal projection of H onto U. If we view the local estimation errors in the calculations as additive noise $e(x,s)$, then

$$\mathcal{V}[f + e] = f + \mathcal{V}[e] \tag{11}$$

Since $\mathcal{V}$ is an orthogonal projection, $\|\mathcal{V}[e]\| \leq \|e\|$. That is, $\mathcal{V}$ removes the component of the noise that is in $U^\perp$ and thereby projects the estimate to a function that is closer to the correct solution f.

This orthogonal projection can be used to increase stability in the psymap associations by reshaping the attractor basins around valid wavelet transforms where small errors result in invalid spectra, which are subsequently removed or greatly reduced by (10). When used in conjunction with a pattern recognition model based on neurotransmitter fields this projection can extend the region of stability.

From the definition of the reproducing kernel (9), we can see that at a fixed position (x_0, s_0) in the spectral manifold, the kernel $K(x_0, s_0, \xi, \eta)$ is zero for values of (ξ, η) where the spectral functions φ_{x_0, s_0} and $\varphi_{\xi, \eta}$ do not overlap. Moreover, K is defined in terms of the wavelets themselves and does not depend on the transformed function f. Consequently, the kernel K can be pre-computed (and encoded in the synaptic connections). The projection defined by (10) can then be evaluated dynamically at each point (x, s).

We have discussed the reproducing kernel only for the case of one-dimensional wavelets $\varphi_{x,s}$. However, the definition (9) is expressed in terms of an inner product, and in general, reproducing kernels only require the mathematical structure of a Hilbert space [37][38].

3.6 A Psymap Architecture

It is possible to design many different types of psymaps using local processing elements. Fig. 8 shows a detailed psymap design that illustrates one possible architecture. Variations of this design can achieve the same or similar functionality.

The double lines in Fig. 8 represent the transfer of data defined on three-dimensional spectral manifolds. The letter $\mathcal{G}$ denotes a general association operator such as the lattice of processing elements shown in Fig. 4A, the letter $\mathcal{V}$ denotes orthogonal projections based on the reproducing kernel, and the vertical box marked $\mathcal{M}$ performs multiplexing operations based upon the mask inputs marked T_S and T_R. Each box labeled $\mathcal{G}$ is trained with a different set of input and output relationships and consequently carries a subscript that identifies it as a unique transform.

The inputs, $\{s_k\}$ and $\{r_k\}$, represent collections of image spectra that arise from either subcortical regions or from the outputs of other Brodmann areas. The integration of data from three separate input images was illustrated in Fig. 5. If $\mathcal{G}_S$ (or $\mathcal{G}_R$) forms the same output association for several members of a given collection of inputs $\{s_i\}$ (or $\{r_i\}$), then the associations will mutually reinforce one another. Consequently, even though single inputs may not be sufficiently strong to bring forth a recollection, multiple inputs will add "context", and their combined effect may surpass a threshold required to evoke the entire memory.

The multiplexer masks allow the system to focus attention on selected regions and to control whether individual Brodmann areas accept new inputs or retain their current contents. Each of the reciprocal Λ-maps, shown in Fig. 8, contains two separate, association elements whose outputs feed into the A and B inputs of the multiplexer box labeled $\mathcal{M}$. A third input, labeled T, shown with a diagonal line, is the control mask.

Let $\alpha(x, y, s)$ be one of the mask signals T_S or T_R. These can control the multiplexer by performing an operation analogous to the *alpha blending* calculation, $(1-\alpha)A + \alpha B$, used for image composition in computer graphics [39]. For points where $\alpha = 0$, the output equals A, and where $\alpha = 1$, the output equals B. Values in between smoothly blend the two images.

When the spectral coefficients are not based on a set of orthogonal functions, the result of an association formed by a real-valued neural network will usually contain small errors that result in invalid spectral representations. Moreover, the masking operation may also result in a spectral function that does not correspond to the transform of an actual image. However, using the reproducing kernel (9) we can project these functions using the linear operator $\mathcal{V}$ given by (10) to the nearest function for which the inverse transform $\mathcal{W}^{-1}$ produces a valid result. This operation is shown in Fig. 8 following the multiplexing operation. Since the results of the orthogonal projections are Q and P, we thereby guarantee that the spectral outputs of a psymap always correspond to valid images.

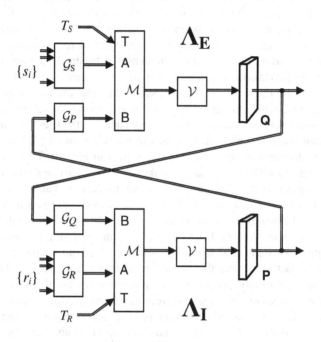

Fig. 8. A detailed psymap implementation illustrates the general pattern of interconnection. $\mathcal{G}$ denotes a neural network association operation, $\mathcal{M}$ denotes the masking and multiplexing operation and $\mathcal{V}$ denotes the orthogonal projection based on the reproducing kernel.

As long as the control signal $\alpha(x, y, s)$ is identically equal to one, in both the exterior and interior Λ-maps, the psymap will ignore its inputs and retain its current stored value. Under these conditions, the psymap is *latched* and the contents of this short-term memory remain fixed on the outputs Q and P.

4 The Psymap Array Model

If each Brodmann area corresponds to a unique psymap, then the collection of all the Brodmann areas in the cerebral cortex constitutes a *psymap array*.

A graphic representation of a psymap array is shown in Fig. 9 where the lines denote the transfer of complete images. To simplify the diagram, we use an "I/O Bus" notation where the horizontal rows along the top of the diagram represent the images, and dots are used to illustrate connections. Sensory inputs are transmitted as computational maps, as are the motor control outputs. Each psymap generates two output images, corresponding to the two component Λ-maps. Physiologically, the two output images arise from the exterior and interior pyramidal layers of the cortex. These are connected figuratively via the I/O bus to other areas. Each psymap can have any number of images as inputs, which are integrated together using topographic alignment.

Each psymap in the psymap array corresponds to a unique Brodmann area. The number and source of the inputs and the number of associations formed will vary between psymaps and even between the internal and external Λ-maps of a single psymap. These variations will affect the computational requirements of the different cortical layers causing them to vary in thickness between Brodmann areas. However, within a single Brodmann area, the number of associations and the type and origin of the inputs will be the same throughout, suggesting that the thicknesses of the layers should be approximately the same. Thus, the psymap array model predicts the observed sharp transitions between Brodmann areas and a uniform thickness within each area [1].

Brodmann mapped the cortical regions of many mammals [40] including those of the monkey in diagrams similar to the one shown in Fig. 10. These drawings can be directly correlated with the psymap array model such as the one shown in Fig. 9.

An assumption underlying some cognitive models is that during the course of evolution, there was a sudden "break" when the CNS developed a new type of feature extraction mechanism that converted images into tokens, instances taken from a small finite set of discrete symbols. It was then able to manipulate these tokens using a new symbolic processing engine that simultaneously evolved during this same time interval. An alternate hypothesis is that this "break" never took place. Instead, additional general-purpose image association processors, computational engines that had already evolved, were replicated to solve problems of increasing difficulty. Most recently, during the last few million years, the already existing psymap array expanded again, adding and integrating new psymaps for language and other aspects of abstract reasoning [41].

4.1 Attention and Control

Inside every CPU, a control unit coordinates the flow of data between the various components. Every area of the neocortex has reciprocal connections with the thalamus [42], which is known to play an important role in consciousness. Based on observed behavior and the types of real-world problems that the CNS must solve, an agile and flexible control mechanism would be based on image subsets or regions. As a first approximation, we can view these image regions as "control masks" that overlay the psymap and regulate the psymap processing and Input/Output. These masks are the T_S or T_R. control images that are shown as inputs to the multiplexer in Fig. 8. The masks blanket the entire extent of the psymap "content" images and consequently can be used to identify an arbitrary region within the image. The portions of the images inside the masked region are active while those outside of the region are quiescent.

Subcortical Inputs and Intracortical Connections

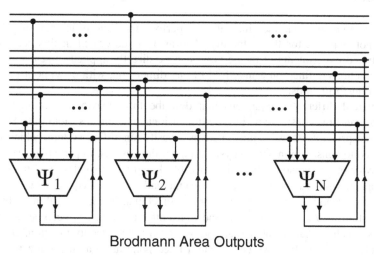

Brodmann Area Outputs

Fig. 9. The psymap array model of the cerebral cortex. Each psymap, Ψ_i, corresponds to a separate Brodmann area. Lines in the diagram correspond to images, and dots represent specific input and output connections.

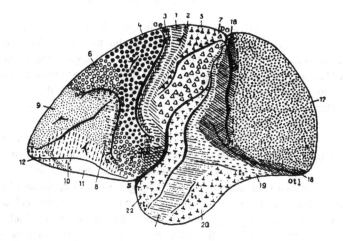

Fig. 10. The Brodmann areas of the monkey cortex. The general principles of operation of a psymap array model can be applied to describe the cerebral cortex of any mammal.

In this way the thalamus can focus attention on the entire image or on some very small detail. Given the variety of computational maps, control masks can be used to direct our attention to a small region in the visual field, a single sound in a word, a particular part of the body surface, or the motion of a single muscle. Time varying control masks can be used to spatially scan a visual image, or temporally scan through the sounds of a sentence.

5 Stability

Frequency domain techniques increase the performance and efficiency of psymaps, but are not required for their stability. The analysis described in this section, and demonstrated experimentally in the next, can be directly applied to scalar or vector valued functions defined on a manifold of any dimension, with or without the use of spectral transformations.

The partial differential equations that describe the behavior of reciprocal-image attractors over time constitute an infinite-dimensional dynamical system [43]. However, some basic principles can be illustrated in a vastly simplified system, which can be generalized to construction and analysis in higher dimensions. In particular, we analyze the behavior of a psymap composed of two black-and-white single-pixel images. This very simple psymap can be expressed by two mutually recurrent equations $y = f(x)$ and $x = g(y)$ where x and y are real numbers. For one-pixel images, the integral in (3) becomes the evaluation of θ at a single point and transmitter functions, τ_i, become real-valued constants which we can as label x_i and y_i. We set σ to the identity function. Using two recursively connected PEs, we can create a fixed point at an arbitrary location (x_0, y_0) with the two Gaussian θ functions

$$
\begin{aligned}
y &= f(x) = y_0 \exp\left(-(x - x_0)^2 / c_x^2\right) \\
x &= g(y) = x_0 \exp\left(-(y - y_0)^2 / c_y^2\right).
\end{aligned}
\tag{13}
$$

It is easy to verify that (x_0, y_0) is a fixed point of (13). The two equations can be combined, $x_{+\Delta t} = g(f(x))$, and differentiated to verify that the fixed point is stable. This can be seen graphically in Fig. 11A. The arrows indicate steps in the recursion (13), which moves toward the stable fixed-point located at the intersection of the two curves. Because the derivatives of f and g are zero at the fixed-point, the convergence is *superlinear*, that is, the rate of convergence increases without bound as the recursion approaches the stable attractor.

In Fig. 11B we show the characteristics of a single-pixel psymap with a total of four processing elements, two processing elements in each of the two Λ-maps. In the neurotransmitter field summation (3), the two PEs generate bimodal functions that contain two bell-shaped Gaussian curves. By setting the ρ_i and τ_i. values of each PE – which correspond to x_0 and y_0 in (13) – we can create two stable reciprocal attractors at arbitrary locations (x_1, y_1) and (x_2, y_2). The attractor basins are shown as shaded areas on the x and y axes in Fig 11B. Within these regions, the psymap will converge toward the stable attractors, as it does in the system shown in Fig 11A.

The curves shown in Fig 11B intersect in five locations; therefore, the system has five fixed-points. Two of these are unstable, but there is a *third stable* fixed-point located near the origin and labeled (x_b, y_b). This point corresponds to a pair of nearly black (zero-valued) reciprocal-images. This stable fixed-point will occur when none of the processing elements "recognizes" the zero input values, which in turn causes the summation of their outputs in both Λ-maps to be nearly zero. When started from initial values outside of the shaded areas shown, the association will "collapse" and the system will converge to the near-zero value (x_b, y_b).

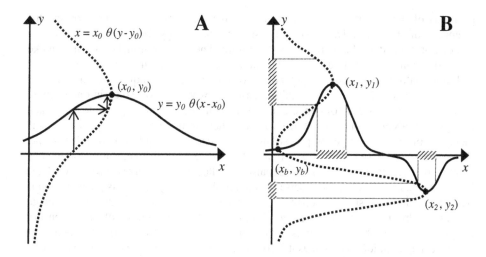

Fig. 11. The recursive equations defined by real-valued Gaussian theta functions (13) create a stable fixed-point attractor at (x_0, y_0) in the system (A). The arrows, which illustrate successive iterations of the recurrence, describe the behavior of a psymap for images near a reciprocal-image attractor. Two stable attractors can be created (B) with two processing elements in each Λ-map. The attractor basins are delineated by shaded regions on the x and y axes.

Because of their non-zero tails, the two Gaussian theta functions will to some extent overlap, which may cause an unwanted "association skew" in the two Λ-maps. This effect can be minimized by setting the standard deviation to a small value. Moreover, bell-shaped theta functions can be created which have compact support, that is, functions that are identically equal to zero outside some finite closed interval.

By decreasing the width of the Gaussian theta functions, new stable attractors can be created by including one additional PE in each of the Λ-maps. As long as their x and y positions do not overlap with an existing attractor, the new attractors can be placed in arbitrary locations. Initial values near any of the reciprocal-image attractors will exhibit a convergence comparable to the system illustrated in Fig. 11A. Physiologically, this implies that we can create a new "memory", that is a new stable attractor, with only two neurons (or glial cells), one in each Λ-map.

The basic principles described for a single-pixel scalar-valued psymap can be generalized to higher dimensions. Many interdependent factors, such as the restriction of the receptor (μ_i) and transmitter (τ_i) functions to local regions, generate complex behavior and confound the analysis. However, while the resulting systems may have any number of periodic or chaotic attractors, it is possible to create local neighborhoods where the psymaps are structurally stable.

We can imagine a large geographic area with attractor basins created by the force of gravity where a frictionless ball would roll into valleys or "energy wells". The attractors correspond to the local minimum in each valley. The ability to specify the ρ_i, μ_i and τ_i functions in (3) in addition to the θ and σ characteristics provides a great deal of flexibility in controlling the potential energy landscape. The patterns control the positions of the valleys, and the theta and sigma functions control the size and the steepness of the sides. Decreasing the size of the energy wells allows the image association capacity to be increased without limit. The smaller attractor size may make the associations more difficult to "find", but once located, the attractors will be stable.

Using multiple psymaps is a potentially very powerful technique since it can serve as a mechanism to trigger a "recognition cascade" where convergence in one or more psymaps causes a cascade of convergence in several others. This is possible since each psymap can accept input images from many other Brodmann areas. Again using the metaphor of gravity and geographical landscapes, each psymap has its own landscape. While we may be searching for a relatively tiny reciprocal-image attractor in one, another psymap may have a large expansive valley where convergence toward the bottom, brings us closer to the precise solution in the first psymap. This phenomenon is well known and well studied in psychology, where verbal "hints" about what an image contains help us locate it. Crossword puzzles are an example of multi-sensory integration, where phonetic sounds are combined with semantic meaning to help find the correct association. As soon as it is located, the association itself is very stable.

6 Experimental Results

A software package written in Java [44] was used to test a variety of psymaps where the data values ranged from single pixels to functions defined on one, two and three dimensional manifolds. In all cases, stable attractors were created analogous to those shown in Fig. 11A and 11B. Consequently, while the results demonstrated are limited to two-dimensional images, similar stable behavior generalizes to spaces of arbitrary dimension which may or may not include frequency components such as those produced by a wavelet transform.

A system with eight stable attractors was constructed from the eight image pairs shown in Fig. 12.

Figures 13, 14 and 15 illustrate the convergence of the psymap to one of the stable reciprocal-image attractors. In each experiment the initial value is shown as the leftmost image Q. In Fig. 13, the photograph of the volcanic eruption obscured by various lines and circles represents a displacement away from the fixed-point attractor corresponding to association A2, but still within its attractor basin. The iterated sequence of images shows the image pair moving toward the stable fixed-point. In Fig. 14, the photograph of "Lena" obscured by the same lines and circles converges instead to the association A4. In Fig. 15, the initial value consists of random circles and line segments added to the Lena test image.

Image pixels effectively correspond to synapses rather than neurons. Consequently for equivalent image resolutions, the processing time and the training time are much less than the corresponding neural network model. In the test case shown, the image resolution was 256×256 but the PEs were spaced 16 pixels apart, resulting in a total of (16+1)×(16+1) PEs. The size of the receptor regions was 64×64 pixels and the transmitter regions were 48×48 pixels. Thus, associations between two 65,536-pixel images were created with 289 PEs where each PE had a localized receptor region of 4,096 pixels.

Fig. 12. The sixteen associated images used to construct a single psymap that contained eight reciprocal-image attractors

The training phase consisted of creating a mesh of PE objects whose overlapping receptor and transmitter regions covered the input and output images respectively. For each image association, a new set of PEs was created and added to the system.

To create a smooth transition between regions the transmitter functions were initialized with a two-dimensional "hill-shape" before scaling by the adjusted output image. The transmitter function, τ_i, of each PE has a significant non-zero multiplier only when an image close to the associated pattern ρ_i is present on the input. Consequently, since each pixel is contained in several overlapping regions, the training phase consists of scaling all the τ_i functions, whose PEs have the same input pattern, so they sum to the desired output pixel values. To calculate the τ_i, all of the two-dimensional hill-shaped initial values are summed pointwise resulting in an image that represents the size and extent of the overlap in the transmitter regions. A scalar coefficient for each pixel is then calculated by dividing the desired image by this summed image. The scalar coefficients are then multiplied pointwise by each of the initial transmitter values, which respond to that input pattern, resulting in a set of transmitter functions whose sum will equal the associated output image. Thus, the initial transmitter regions can have arbitrary size and shape as long as they collectively cover the output manifold.

All of the images used had a resolution of 256×256 pixels, but the processing element grid was only 17x17. Since the computations are primarily inner products and summations, they execute quickly and are suitable for parallel implementations. For the results shown, the set-up time was less than 10s and the run time for each of the tests was approximately 30s on a single processor 2 GHz personal computer.

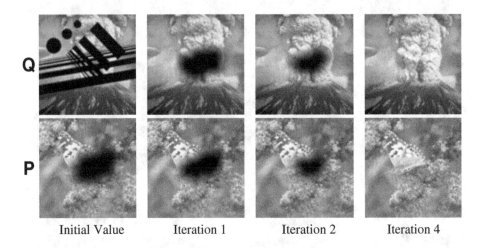

| Initial Value | Iteration 1 | Iteration 2 | Iteration 4 |

Fig. 13. An initial value that converges to the association A2 reciprocal-image attractor

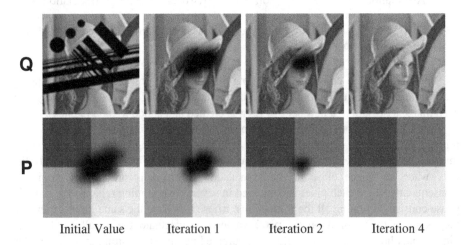

| Initial Value | Iteration 1 | Iteration 2 | Iteration 4 |

Fig. 14. An initial value that converges to association A4. Creating an association with a square "barcode" image pattern helps assist in recognition and classification.

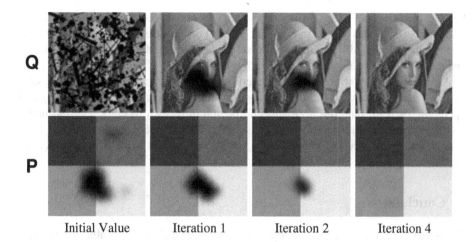

| | Initial Value | Iteration 1 | Iteration 2 | Iteration 4 |

Fig. 15. An initial value obscured by numerous random lines and circles that also "collapses" locally in some areas, but is still within the global attactor basin of association A4

The training and execution times are a significant improvement over common neural-network algorithms for several reasons related to the fundamental differences in the computational models. During the set-up or training phase, time-consuming back propagation methods are not required since learning takes place by adding cells which recognize an input image, and directly generate the associated output image. In the neurotransmitter-field model, neither the training phase nor the testing phase uses a slow (and often error prone) minimization algorithm. Moreover, as was discussed in section 5, the convergence rate near the fixed-point attractor is superlinear. The execution time is also faster since an action potential does not correspond to the value of a single pixel, but rather acts as a signal to all of the synapses in an axonal tree which define the image values over an entire region.

The rate and nature of the convergence shown in Figs 13-15 can be modified by changing the standard deviation of the theta functions, as well as the slope and inflection point of the sigma functions. Small values of the standard deviation for θ prevent interference between image associations. For the results illustrated, a modified hyperbolic tangent function was used for sigma transfer function with the inflection point at 0.15 and the slope at the inflection point equal to 40. If the inflection point for σ were set to a relatively large value such as 0.75, then more than 75% of the pixels in each PE receptor region μ_i would have to match (approximately) to produce a coefficient for τ_i of greater than 0.5. In this case, the initial values used in the test cases would diverge to all black images. Additional experimental results with automatically generated HTML reports, which document the parameter settings and computed outputs, are available online [45].

Locally, many of the characteristics of single-pixel psymaps, illustrated in Fig 11, also apply to psymaps defined on multi-dimensional manifolds. For example, in the associations A3 and A4, both of the P images are square "barcode" images. However, because the grayscale values differ, the associations remain separated as they do for the fixed-points (x_1,y_1) and (x_2,y_2) in Fig. 11B. The test results also show how in some local regions, there are insufficient matching pixel values, causing the association to collapse and move toward the all black (x_b,y_b) fixed-point.

A central concept, which all three experiments help illustrate, is that global system behavior can be generated from the purely local action of neurons. In both training and operation, each PE only has access to information in its own receptor and transmitter regions. However, the neighboring cells interact to re-establish the missing local details. The recurrence in the overlapping regions of support make it possible for the local connections to establish overall global associations.

7 Conclusions

The stability of the computation results demonstrates that it is possible to associate arbitrary images rather than discrete bit patterns in the construction of a general-purpose symbolic processor. However, additional study is required in several areas. From the purely mathematical point of view, psymaps and psymap arrays are interesting but complicated objects. Images are defined in infinite-dimensional function spaces, and determining the characteristics of the attractor basins, as well as the behavior of the system when external (sensory) input images are constantly changing, is important but difficult. From a computational perspective, the effectiveness of local spectral representations and the orthogonal projection using the reproducing kernel must be measured and analyzed. Finally, and most important, the validity of the model as a scientific explanation needs to be established or refuted. The theory presented contains several interrelated propositions, which include the neurotransmitter cloud hypothesis and the existence of reciprocal-image attractors in the cerebral cortex. The current lack of contradicting scientific facts cannot be construed as a strong supporting argument, and it seems likely that targeted biological experiments will be required.

The psymap is a new type of image centric processing architecture, which combines several design elements including recurrence of locally connected processing elements, spectral transformations, and an orthogonal projection based on the reproducing kernel. Compared to a neuron action-potential model, a neurotransmitter-field model presents a broader view of natural intelligence. It allows the chemical reactions that take place in many types of cells, including neuroglia, to be incorporated into a general framework of memory and computation. Direct associations between images are shown to be stable, and therefore the images themselves can act as symbols. The psymap array model has many correlations with the neurological organization of the cerebral cortex and may help provide insight into the nature of consciousness and cognition.

Acknowledgments. The author would like to thank Professor M. Tuceryan for his support and helpful comments. This document contains material that is patent pending by General Manifolds LLC, http://www.gmanif.com/ip.

References

1. Brodmann, K.: Vergleichende Lokalisationslehre der Grosshirnrinde in ihren Prinzipien dargestellt auf Grund des Zellenbaues. Barth, Leipzig (1909)
2. Kandel, E.R., Schwartz, J.H., Jessell, T.M.: Principles of Neural Science, 4th edn. McGraw-Hill, New York (2000)
3. Garoutte, B.: Survey of Functional Neuroanatomy, 3rd edn. Mill Valley Medical, California (1994)
4. Spivak, M.: A Comprehensive Introduction to Differential Geometry, 3rd edn. Publish or Perish, Wilmington (1979)
5. Schutz, B.: Geometric Methods of Mathematical Physics. Cambridge University Press, Cambridge (1980)
6. Knudsen, E.I., du Lac, S., Esterly, S.D.: Computational Maps in the Brain. Ann. Rev. of Neuroscience 10, 41–65 (1987)
7. Greer, D.S.: A Unified System of Computational Manifolds. Tech. Rep. TR-CIS-0602-03, Dept. of Comp. and Info. Sci., IUPUI, Indianapolis (2003)
8. Monad, J.: Chance and Necessity: An Essay on the Natural Philosophy of Modern Biology (Trans. A. Wainhouse). Knopf, New York (1971)
9. Mallat, S.: A Wavelet Tour of Signal Processing, 2nd edn. Academic Press, San Diego (1999)
10. Wang, Y.: The Cognitive Processes of Abstraction and Formal Inferences. In: Proc. 4th IEEE International Conf. on Cognitive Informatics, Irvine, California (2005)
11. Wang, Y.: The Theoretical Framework of Cognitive Informatics. International J. of Cognitive Informatics and Natural Intelligence 1(1), 1–27 (2007)
12. Yao, Y.Y.: Concept Formation and Learning: a Cognitive Informatics Perspective. In: Proc. 3rd IEEE International Conf. on Cognitive Informatics, Victoria, Canada (2004)
13. Prince, V., Lafourcade, M.: Mixing Semantic Networks and Conceptual Vectors Application to Hyperonymy. IEEE Trans. Systems, Man and Cybernetics, C 36, 152–160 (2006)
14. Willshaw, D.J., Buneman, O.P., Longuet-Higgins, H.C.: Non-holographic Associative Memory. Nature 222, 960–962 (1969)
15. Kosko, B.: Bidirectional Associative Memories. IEEE Trans. Systems, Man and Cybernetics 18, 49–60 (1988)
16. Xu, Z., Leung, Y., He, X.: Asymmetrical Bidirectional Associative Memories. IEEE Trans. Systems, Man, and Cybernetics 24, 1558–1564 (1994)
17. Chua, L.O., Yang, L.: Cellular Neural Networks: Theory. IEEE Trans. Circuits and Systems 35(10), 1257–1272 (1988)
18. Zhang, Q., Benveniste, A.: Wavelet Networks. IEEE Trans. Neural Networks 3(6), 889–898 (1992)
19. Thuillard, M.: A Review of Wavelet Networks, Wavenets, Fuzzy Wavenets and their Applications. In: Advances in Computational Intelligence and Learning: Methods and Applications. Kluwer, Deventer (2002)
20. Iyengar, S.S., Cho, E.C., Phoha, V.V.: Foundations of Wavelet Networks and Applications. CRC, Boca Raton (2002)
21. Beurle, R.L.: Properties of a Mass of Cells Capable of Regenerating Pulses. Philosophical Trans. of the Royal Society. Series B 240, 55–94 (1956)
22. Griffith, J.S.: A Field Theory of Neural Nets: I: Derivation of Field Equations. Bulletin of Mathematical Biophysics 25, 111–120 (1963)

23. Griffith, J.S.: A Field Theory of Neural Nets: II. Properties of the Field Equations. Bulletin of Mathematical Biophysics 27, 187–195 (1965)
24. Amari, S.: Dynamics of Pattern Formation in the Lateral-Inhibition Type Neural Fields. Biological Cybernetics 27, 77–87 (1977)
25. Ermentrout, B.: Neural Networks as Spatio-temporal Pattern-Forming Systems. Reports on Progress in Physics 61, 353–430 (1998)
26. Vogels, T.P., Rajan, K., Abbott, L.F.: Neural Network Dynamics. Annual Rev. Neuroscience 28, 357–376 (2005)
27. Sarnat, H.B., Netsky, M.G.: Evolution of the Nervous System, 2nd edn. Oxford (1981)
28. Widrow, B., Hoff, M.E.: Adaptive Switching Circuits. WESCON Convention Record, pp. 96–104. IRE, New York (1960)
29. Royden, H.L.: Real Analysis, 3rd edn. Prentice-Hall, Englewood Cliffs (1988)
30. Greer, D.S.: Neurotransmitter Fields. In: Proc. of International Conf. on Artificial Neural Networks, Porto, Portugal (2007)
31. Bishop, C.M.: Neural Networks for Pattern Recognition. Oxford University Press, Oxford (1995)
32. Thompson, J.M.T., Stewart, H.B.: Nonlinear Dynamics and Chaos. Wiley, Hoboken (2002)
33. Nevatia, R.: Machine Perception. Prentice-Hall, Englewood Cliffs (1982)
34. Churchland, P., Sejnowski, T.J.: The Computational Brain. MIT Press, Cambridge (1994)
35. Daubechies, I.: Ten Lectures on Wavelets. Society for Industrial and Applied Mathematics, Philadelphia (1992)
36. Addison, P.S.: The Illustrated Wavelet Transform Handbook. Institute of Physics Publishing, Bristol (2002)
37. Aronszajn, N.: Theory of Reproducing Kernels. Trans. of the American Mathematical Society 68(3), 337–404 (1950)
38. Saitoh, S.: Integral Transforms, Reproducing Kernels and their Applications. Addison Wesley Longman, Essex (1997)
39. Thompson, K.: Alpha Blending. In: Glassner, A. (ed.) Graphics Gems, pp. 210–211. Academic Press, Cambridge (1990)
40. von Economo, G.N., Koskinas, G.N.: The Cytoarchitectonics of the Adult Human Cortex (trans. H. L. Seldon). Springer, Heidelberg (1925)
41. Greer, D.S.: The Computational Manifold Approach to Consciousness and Symbolic Processing in the Cerebral Cortex. In: Proc. 7th IEEE International Conference on Cognitive Informatics (2008)
42. Behrens, T.E.J., Johansen-Berg, W.H.M.W., Smith, S.M., Wheeler-Kingshott, C.A.M., Boulby, P.A., et al.: Non-invasive Mapping of Connections between Human Thalamus and Cortex Using Diffusion Imaging. Nature Neuroscience 6(7), 750–757 (2003)
43. Robinson, J.C.: Infinite-Dimensional Dynamical Systems: An Introduction to Dissipative Parabolic PDEs and the Theory of Global Attractors. Cambridge University Press, Cambridge (2001)
44. Greer, D.S.: Sapphire 0.4 Implementation Notes. Tech. Rep. TR-CIS-0714-08, Dept. of Comp. and Info. Sci., IUPUI, Indianapolis (2008)
45. Computational Results, http://www.gmanif.com/results.html

Knowledge Reduction of Covering Approximation Space*

Jun Hu[1,2] and GuoYin Wang[2]

[1] School of Electronic Engineering, XiDian University,
Xi'an, Shaanxi, 710071, P.R. China
[2] Institute of Computer Science and Technology,
Chongqing University of Posts and Telecommunications,
Chongqing, 400065, P.R. China
{hujun,wanggy}@cqupt.edu.cn

Abstract. Knowledge reduction is a key issue in data mining. In order to simplify the covering approximation space and mining rules from it, Zhu proposed a reduction of covering approximation space which does not rely on any prior given concept or decision. Unfortunately, it could only reduce absolutely redundant knowledge. To reduce relatively redundant knowledge with respect to a given concept or decision, the problem of relative reduction is studied in this paper. The condition in which an element of a covering is relatively reducible is discussed. By deleting all relatively reducible elements of a covering approximation space, one can get the relative reduction of the original covering approximation space. Moreover, one can find that the covering lower and upper approximations in the reduced space are the same as in the original covering space. That is to say, it does not decrease the classification ability of a covering approximation space to reduce the relatively reducible elements in it. In addition, combining absolute reduction and relative reduction, an algorithm for knowledge reduction of covering approximation space is developed. It can reduce not only absolutely redundant knowledge, but also relatively redundant knowledge. It is significant for the following-up steps of data mining.

Keywords: Covering, reduction, granule, rough sets.

1 Introduction

It is a key issue in artificial intelligence to process vague information. Fuzzy logic was proposed by Zadeh [16], which is different from classical set theory. It

---

* This research is supported by the National Natural Science Foundation of P.R. China (No.60573068, No.60773113), Natural Science Foundation of Chongqing (No.2005BA2003, No.2008BA2017), Natural Science Foundation of Chongqing University of Posts and Telecommunications (A2006-56), and Science & Technology Research Program of the Municipal Education Committee of Chongqing (No.KJ060517).

M.L. Gavrilova et al. (Eds.): Trans. on Comput. Sci. V, LNCS 5540, pp. 69–80, 2009.

allows an object to be a partial member of a set. However, it needs additional information about the data, namely membership function. Rough set theory was proposed by Pawlak [10]. The key idea of rough set is to approximate an arbitrary set by two definable sets of a universe, called the lower approximation and upper approximation. For its ability in processing information featured with imprecise, inconsistent and incomplete, it has been applied successfully in many fields in the past years, such as process control, economics, medical diagnosis, biochemistry, environmental science, biology, chemistry psychology, conflict analysis, emotion recognition, video retrieval, and so on [9][14][15][18][23].

Pawlak's rough set theory is based on an indiscernibility relation, which is an equivalence relation. An indiscernibility relation determines a partition, and each element of the partition is called an equivalence class. Furthermore, if a subset of the universe can be represented by a union of some equivalence classes, then it is a definable set. All these definable sets form an approximation space. An arbitrary subset of the universe can be approximated by two definable sets, called lower approximation and upper approximation. The lower approximation is the maximal definable set included in it, while the upper approximation is the minimal definable set including it.

Unfortunately, it might be impossible or cost too much to acquire an indiscernibility relation in some real life cases. Thus, it is necessary and important to extend rough set theory into general cases. As we know, given an equivalence relation on a universe, then we can get a partition on it, and vice versa. However, a covering on a universe has not such a one to one relation with a general relation on the universe. So, some researchers studied extended rough set models from two different views, general relation based rough set model [4][5][6][11][13] and covering based rough set model [1][2][3][7][8][12][17][19][20][21]. The generalized rough set theory based on general relation has been studied by many researchers, whereas there are still some key issues needed to be studied for the covering based rough set theory, such as the reduction of covering approximation space addressed in this paper.

Although the concept of covering approximation space (CAS) was proposed by Zakowski in 1983 [17], it did not get much attention until recent years. In 1998, Bonikowski studied the rough approximations in covering [1]. Latter, other three different definitions for rough approximations were developed [8][12][24]. Then, the relationship of these four types of covering rough set models was analyzed [22]. From the above works, we can see that they all developed the covering based rough set theory by constructive approach. In order to study the structure of covering based rough set algebra, the axiomatic systems for the covering lower and upper approximation operations were constructed [20][21]. Two rough entropies were also developed for measuring the uncertainty of a covering approximation space [2][3], which provided tools for quantitatively analyzing the approximation ability of a given covering approximation space. Moreover, Zhu discussed the reduction of a covering approximation space and the condition in which two coverings could generate the same covering lower or upper approximations for any concept [19][20].

In data mining, it is an important issue to reduce redundant information and get concise rules. However, the reduction algorithm proposed by Zhu can reduce absolutely redundant knowledge only. That is to say, the reduced covering has the same approximation ability as the original covering, or the reduced covering and the original covering could generate the same covering lower and upper approximations for any concept. In order to acquire the maximally generalized decision rules, a reduction is developed in this paper which can reduce relatively redundant knowledge with respect to a given concept or a given decision. To distinguish it from the reduction defined by Zhu, we call it relative reduction, and the reduction defined by Zhu absolute reduction. Based on this definition of relative reduction and absolute reduction, a diagram of knowledge reduction and its algorithm are developed. It can reduce not only absolutely redundant knowledge, but also relatively redundant knowledge. As a result, it can simplify a covering approximation space, and decrease the computing complexity of acquiring knowledge from it.

We organize the remainder of this paper as follows. Some basic concepts about covering approximation space are introduced in section 2. Section 3 discusses the absolute reduction of covering approximation space. The definition of relative reduction of covering approximation space and its properties are studied in section 4. A diagram for knowledge reduction of covering approximation space and its algorithm are developed in section 5. Section 6 concludes the paper.

2   Basic Concept of Covering Approximation Space

For the convenience of following discussion, some basic concepts of covering approximation space are introduced at first.

Definition 1. *Let U be a universe of discourse, C be a family of subsets of U. If none subset in C is empty, and $\cup C = U$, C is called a covering of U.*

In the following discussion, unless stated to be the contrary, the universe and the covering are considered to be finite.

Definition 2. *Let U be a universe of discourse, C be a covering of U. We call the ordered pair $< U, C >$ a covering approximation space, or covering space for short.*

Definition 3. *Let $< U, C >$ be a covering approximation space, for an element $x(x \in U)$, then the set family $Md(x) = \{K \in C | x \in K \wedge \forall_{S \in C}(x \in S \wedge S \subseteq K \Rightarrow K = S)\}$ is called the minimal description of x.*

Concept approximation of covering approximation space, namely how to define the covering lower and upper approximations of a given concept, is a key problem for acquiring knowledge from it. Until now, four types of covering based rough set models have been developed.

Definition 4. *Let $< U, C >$ be a covering approximation space, for a set $X(X \subseteq U)$, the covering lower and upper approximations of the set X are defined as follows:*

$$CL(X) = \cup \{K | K \in C \wedge K \subseteq X\} \qquad (1)$$

$$FH(X) = CL(X) \cup \{Md(x) | x \in X - CL(X)\} \qquad (2)$$

$$SH(X) = \cup \{K | K \in C \wedge K \cap X \neq \phi\} \qquad (3)$$

$$TH(X) = \cup \{Md(x) | x \in X\} \qquad (4)$$

$$RH(X) = CL(X) \cup \{K | K \in C \wedge K \cap (X - CL(X)) \neq \phi\} \qquad (5)$$

Where CL is the covering lower approximation, and FH, SH, TH and RH are the first, the second, the third and the fourth types of covering upper approximation, respectively.

For the convenience of discussion, we use $C_*(X)$ to represent the covering lower approximation, and $C^*(X)$ the covering upper approximation, which could be the first, the second, the third or the fourth types of covering upper approximation. If $C_*(X) = C^*(X)$ then X is called relatively exact with respect to C, otherwise, X is called relatively inexact with respect to C.

Generally speaking, given two different coverings and a subset (concept) of a universe, according to Definition 4, two different covering lower approximations, as well as two different covering upper approximations, of a concept will be induced. If two different coverings could generate the same covering lower and upper approximations for any concept, then we say these two coverings are equal in knowledge, or they have the same approximation ability. Moreover, if the approximation ability of a covering space will not decrease after deleting an element from it, then we say this element is reducible in this covering. In the following section, the absolute reduction will be discussed.

3   Absolute Reduction of Covering Approximation Space

In order to find the condition in which two different coverings could generate the same covering lower and upper approximations, Zhu proposed the definition of reducible element and irreducible element of a covering. By deleting all reducible elements in a covering, one can get a minimal covering with the same approximation ability as the original one for any concept. In this section, the absolute reduction of covering and its important properties are introduced.

Definition 5. *Let C be a covering of a universe U, K be an element of C. If K is a union of some sets in $C - \{K\}$, we say K is a reducible element of C, otherwise an irreducible element of C.*

Definition 6. *Let C be a covering of U. If every element of C is an irreducible element, we say C is irreducible, otherwise, C is reducible.*

Proposition 1. *Let C be a covering of U, K be a reducible element of C, and K_1 be another element of C, then K_1 is a reducible element of if and only if it is a reducible element of $C - \{K\}$.*

Definition 7. *For a covering C of a universe U, the new irreducible covering through eliminating the reducible elements of C is called the reduction of C, denoted by $reduct(C)$, and this reduction is unique.*

In order to distinguish this reduction from the reduction defined in the following section, we call it absolute reduction of covering approximation space, or absolute reduction for short.

Theorem 1. *Let C be a covering of U, $reduct(C)$ be the absolute reduction of C, then C and $reduct(C)$ generate the same covering lower approximations.*

Theorem 2. *Let C be a covering of U, $reduct(C)$ be the absolute reduction of C, then C and $reduct(C)$ generate the same first type of covering upper approximations.*

Theorem 3. *Let C be a covering of U, $reduct(C)$ be the absolute reduction of C, then C and $reduct(C)$ generate the same third type of covering upper approximations.*

However, a covering and its reduction may generate different results for the second and the fourth types of covering upper approximations. So, we will only discuss the first and the third types of covering based rough set models in the following sections.

From the above introduction, we can find that the absolute reduction of a covering will generate the same covering lower approximation, the same first and third types of covering upper approximations as the original covering. That is to say, it can simplify a covering approximation space by reducing all the reducible elements of it, but does not decrease its approximation ability.

4 Relative Reduction of Covering Approximation Space

According to the above introduction, one can find that any concept of a universe has the same covering lower and upper approximations in a covering and its absolute reduction. That is to say, the absolute reduction can reduce redundant knowledge not relying on a given concept. In many cases, we want to get a minimal covering with respect to a given concept. Then, for a given concept, is there still any other redundant knowledge in a covering approximation space besides the absolutely redundant knowledge? What is the condition in which two coverings could generate the same covering lower and upper approximations for a given concept? In this section, these problems are discussed.

Definition 8. *Let C be a covering of U, X be a subset of U, and K be an element of C. If there exists another element K' of C such that $K \subseteq K' \subseteq X$, then we say K is a relatively reducible element of C with respect to X, or a relatively reducible element for short, otherwise, a relatively irreducible element.*

Definition 9. *Let C be a covering of U, X be a subset of U. If every element of C is relatively irreducible, then we say C is irreducible with respect to X, otherwise, C is reducible with respect to X.*

Proposition 2. *Let C be a covering of U, X be a subset of U. If K is a relatively reducible element of C with respect to X, $C - \{K\}$ is still a covering of U.*

Proposition 3. *Let C be a covering of U, X be a subset of U. If K is a reducible element of C with respect to X, and K' is another element of C, then K' is a relatively reducible element of C with respect to X if and only if it is a relatively reducible element of $C - \{K\}$ with respect to X.*

Proof. If K' is a relatively reducible element of $C - \{K\}$ with respect to X, then there should be an element $K_1 \in C - \{K, K'\}$ such that $K' \subseteq K_1 \subseteq X$. Obviously, $K_1 \in C - \{K\}$, so K' is also a reducible element of C with respect to X. On the other hand, if K' is a relatively reducible element of C with respect to X, then there should be an element $K_1 \in C - \{K'\}$ such that $K' \subseteq K_1 \subseteq X$. If $K_1 \neq K$, then $K_1 \in C - \{K, K'\}$. So, K' is a relatively reducible element of $C - \{K\}$ with respect to X. Otherwise, if $K_1 = K$, since K is a reducible element of C with respect to X, then there should be an element $K_2 \in C - \{K\}$ such that $K \subseteq K_2 \subseteq X$. Thus, $K' \subseteq K_2 \subseteq X$, so K' is also a relatively reducible element of $C - \{K\}$ with respect to X.   □

Proposition 2 guarantees that it is still a covering after deleting a relatively reducible element of a covering. Proposition 3 indicates that it will not generate any new relatively reducible elements or make any other originally relatively reducible elements become relatively irreducible after deleting a relatively reducible element of a covering.

Definition 10. *Let C be a covering of U, X be a subset of U. The new irreducible covering through eliminating all relatively reducible elements is called the relative reduction of C with respect to X, denoted by $reduct_X(C)$.*

Theorem 4. *Let C be a covering of U, X be a subset of U, then there is only one relative reduction of C with respect to X.*

Proposition 4. *Let C be a covering of U, X be a subset of U. If $K(K \in C)$ is a relatively reducible element, then the covering lower approximations of X generated by C and $C - \{K\}$ are the same.*

Proof. Suppose the covering lower approximations of X generated by C and $C - \{K\}$ are X_*^1 and X_*^2 respectively. From the definition of the covering lower approximation, it is evident $X_*^2 \subseteq X_*^1 \subseteq X$. If $x \in X_*^1$, then $\exists_{K_2 \in C}(K_2 \subseteq X)$. If $K_2 \neq K$, then $x \in X_*^2$. If $K_2 = K$, since K is a relatively reducible element with respect to X, there should be an element $K_1(K_1 \in C - \{K\})$ such that $K \subseteq K_1 \subseteq X$, then $x \in X_*^2$ too. Therefore, $X_*^1 \subseteq X_*^2$. So $X_*^1 = X_*^2$, namely Proposition 4 holds.   □

Corollary 1. *Let C be a covering of U, X be a subset of U, and $reduct_X(C)$ be a relative reduction of C with respect to X, then the covering lower approximations of X generated by C and $reduct_X(C)$ are the same.*

Proposition 5. *Let C be a covering of U, X be a subset of U. If $K(K \in C)$ is a relatively reducible element, then the covering upper approximations of X generated by C and $C - \{K\}$ are the same.*

Proof. Since K is a relatively reducible element with respect to X, there should be an element $K_1(K_1 \in C - \{K\})$ such that $K \subseteq K_1 \subseteq X$. Therefore, we have $K \subseteq K_1 \subseteq C_*(X) \subseteq C^*(X)$ no matter which type of covering upper approximation we choose. So, it does not reduce the covering upper approximation to eliminate a relatively reducible element of a covering, namely Proposition 5 holds. □

Corollary 2. *Let C be a covering of U, X be a subset of U, and $reduct_X(C)$ be a relative reduction of C with respect to X, then the covering upper approximations of X generated by C and $reduct_X(C)$ are the same.*

Theorem 5. *Let C be a covering of U, X be a subset of U, and $reduct_X(C)$ be a relative reduction of C with respect to X, then C and $reduct_X(C)$ generate the same covering lower and upper approximations of X.*

Example 1. Let $U = \{x_1, x_2, x_3, x_4\}$. $C = \{K_1, K_2, K_3\}$, where $K_1 = \{x_1\}$, $K_2 = \{x_1, x_2\}$, $K_3 = \{x_2, x_3, x_4\}$. $X = \{x_1, x_2, x_3\}$. Since $K_1 \subseteq K_2 \subseteq X$, K_1 is a relatively reducible element of C with respect to X. So, $reduct_X(C) = \{K_2, K_3\}$. Then, $CL(X) = \{x_1, x_2\}$, $FH(X) = \{x_1, x_2, x_3, x_4\}$ and $TH(X) = \{x_1, x_2, x_3, x_4\}$ in C or $reduct_X(C)$.

From the above discussion, we can find that the relatively reducible elements of a covering are also redundant knowledge. They are different from the absolutely reducible elements since they rely on a given concept. To sum up, the relative reduction of a covering with respect to a given concept may has not the same approximation ability as the original covering for any concept, but it has the same approximation ability for a given concept.

5 Knowledge Reduction of Covering Approximation Space

The purpose of data mining is to acquire knowledge (rules) from large data sets. In a general way, the more concise the rules are, the more useful the rules will be. Therefore, knowledge reduction is a key step of knowledge acquisition. In this section, based on the above results, knowledge reduction of covering approximation space is discussed.

Definition 11. *Let U be a universe of discourse, P be a family of nonempty subsets of U. If $\forall_{p_1, p_2 \in P}(p_1 \neq p_2 \Rightarrow p_1 \cap p_2 = \phi)$, and*

$$\bigcup_{p_i \in P} p_i = U$$

then P is called a partition of U, or a decision.

Definition 12. *Let* $< U, C >$ *be a covering approximation space, and* P *be a decision. For an element* $K(K \in C)$*, if* $\exists_{p_i \in P}(K \subseteq p_i)$*, we call* K *a consistent granule with respect to* P*, or a consistent granule for short, otherwise, an inconsistent granule.*

In knowledge acquisition, rules are often used to represent knowledge. For example, $p \to q$ is a rule, and $|p \cap q|/|U|$ (where $|.|$ is the cardinality of a set) is called its support degree. Generally speaking, a rule with higher support degree would be more useful than a lower one.

Let $< U, C >$ be a covering approximation space, P be a decision. For two granules $K, K'(K, K' \in C)$, if $\exists_{p_i \in P}(K \subseteq K' \subseteq p_i)$, then both K and K' are consistent granules according to Definition 12. However, we say rule $K' \to p_i$ is more useful than rule $K \to p_i$ for its higher support degree, while rule $K \to p_i$ is less useful. Therefore, for a consistent granule, if there is another consistent granule including it, then the rule generated by it is reducible. Obviously, a consistent granule with lower support degree is a relatively reducible element discussed in section 4.

In the other hand, for a granule $K(K \in C)$, suppose K is an inconsistent granule, and

$$K = \bigcup_{k_i \in C - \{K\}} k_i$$

where k_i is absolute irreducible element of C. For a granule k_i and an element x of it, if $\exists_{p_i \in P}(k_i \subseteq p_i)$, then k_i is a consistent granule, and x can be classified into p_i definitely according to rule $k_i \to p_i$. Otherwise, k_i is an inconsistent granule. We know that $k_i \subseteq K$ and K is not in $Md(x)$. If $x \in p_i$, then x can be classified into p_i approximately by rule $k_i \to p_i$. Therefore, for an inconsistent granule, if it is a union of some other granules, the rule generated by it will also be reducible. Apparently, an inconsistent granule constructed by a union of some other granules is an absolute redundant element introduced in section 3.

Definition 13. *Let* U *be a universe,* C *be a covering of* U*, and* P *be a decision. If* $K(K \in C)$ *is relatively reducible with respect to any* $p_i(p_i \in P)$*, then we say* K *is relatively reducible with respect to* P*. Otherwise,* K *is relatively irreducible with respect to* P*.*

Definition 14. *Let* U *be a universe,* C *be a covering of* U*, and* P *be a decision. If every element of* C *is relatively irreducible with respect to* P*, we say* C *is relatively irreducible with respect to* P*. Otherwise,* C *is relatively reducible with respect to* P*.*

Definition 15. *Let* U *be a universe,* C *be a covering of* U*, and* P *be a decision. If any element of* C *is relatively reducible with respect to* P*, then delete it from* C *until* C *is relatively irreducible with respect to* P*. Thus, one will get a relative reduction of* C *with respect to* P*, denoted by* $reduct_P(C)$*.*

Definition 16. *Let* U *be a universe,* C *be a covering of* U*, and* P *be a decision. The positive region and boundary region of* C *with respect to* P *are defined as follows:*

$$Pos_C(P) = \bigcup_{p_i \in P} C_*(p_i) \qquad (6)$$

$$BN_C(P) = \bigcup_{p_i \in P} C^*(p_i) - \bigcup_{p_i \in P} C_*(p_i) \qquad (7)$$

Where $Pos_C(P)$ is a set composed by all elements which could be classified definitely in $< U, C >$, and $BN_C(P)$ is the complement of $Pos_C(P)$, that is to say, $BN_C(P) = U - Pos_C(P)$. Since the definition of covering lower approximation in four types of covering based rough set models is the same, the positive region and boundary region of C with respect to P are also the same in these models.

Theorem 6. *Let C be a covering of U, P be a decision, and $reduct(C)$ be the absolute reduction of C, then $reduct(C)$ and C will generate the same positive regions and boundary regions.*

Theorem 7. *Let C be a covering of U, P be a decision, and $reduct_P(C)$ be the relative reduction of C, then $reduct_P(C)$ and C will generate the same positive regions and boundary regions.*

Theorem 6 and Theorem 7 show that both absolute reduction and relative reduction do not reduce the positive region, or change the boundary region, namely the classification ability is not reduced.

From the above discussion, we could develop the diagram for knowledge reduction of covering approximation space shown in Fig.1.

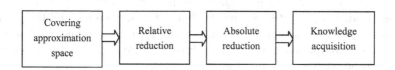

Fig. 1. Knowledge reduction of covering approximation space

The algorithm for knowledge reduction of covering approximation space is as follows:

Algorithm 1 (Knowledge Reduction of Covering Approximation Space)
Input: An covering approximation space $< U, C >$, and a decision P.
Output: $reduct(C|P)$.
Step 1: Let $reduct(C|P) = C$.
Step 2: If $C \neq \phi$, then choose an element $K(K \in C)$, and let $C = C - \{K\}$. Otherwise, let $T = reduct(C|P)$, and go to Step 4.
Step 3: If K is a relatively reducible element of $reduct(C|P)$ with respect to P, then let $reduct(C|P) = reduct(C|P) - \{K\}$. Go to Step 2.
Step 4: If $T \neq \phi$, then choose an element $K(K \in T)$, and let $T = T - \{K\}$. Otherwise, go to Step 6.

Step 5: If K is an absolutely reducible element of $reduct(C|P)$, then let $reduct$ $(C|P) = reduct(C|P) - \{K\}$. Go to Step 4.

Step 6: Output $reduct(C|P)$.

To illustrate the Algorithm 1, let us look at the following example.

Example 2. Suppose the following incomplete information table is given, where x_1, ..., x_5 are the available objects, c_1, ..., c_3 are three attributes and d is a decision attribute classifying objects either to the set ϕ or to the set ψ. We use * to denote the unknown value.

Table 1. An incomplete information table

A	x_1	x_2	x_3	x_4	x_5
c_1	1	*	2	2	*
c_2	1	1	1	2	3
c_3	1	*	2	*	1
d	ϕ	ϕ	φ	φ	φ

Generally, we can use general relation based rough set extensions to address this problem, such as tolerance relation, similarity relation, etc. Here, we use another method to copy this problem, namely covering based rough set theory.

By attribute c_1, two granules can be generated, one is $K_1 = \{x_1\}$, the other is $K_3 = \{x_3, x_4\}$. The semantics of K_1 and K_3 are $c_1 = 1$ and $c_1 = 2$ respectively. Analogously, we can get other granules by attribute c_2, c_3 and d. We call these granules generated by one attribute atomic granules, and the granules built recursively from atomic granules by using logic connections, such as $\neg$, $\wedge$, $\vee$, $\rightarrow$ and $\leftrightarrow$, compound granules. For simplicity, we only discuss the covering approximation spaces made up of atomic granules.

Using all atomic granules generated by c_1, c_2 and c_3, we can get a covering approximation space $< U, C >$, where $U = \{x_1, x_2, x_3, x_4, x_5\}$, $C = \{K_1, K_2, K_3, K_4, K_5, K_6, K_7\}$, $K_1 = \{x_1\}$, $K_2 = \{x_1, x_2, x_3\}$, $K_3 = \{x_3, x_4\}$, $K_4 = \{x_3\}$, $K_5 = \{x_4\}$, $K_6 = \{x_5\}$, $K_7 = \{x_1, x_5\}$. $P = \{p_1, p_2\}$, where $p_1 = \{x_1, x_2\}$, $p_2 = \{x_3, x_4, x_5\}$. Since $K_4 \subseteq K_3 \subseteq p_2$, K_4 is relatively reducible with respect to P, and K_5 is also relatively reducible with respect to P for $K_5 \subseteq K_3 \subseteq p_2$. Moreover, because $K_7 = K_1 \cup K_6$, K_7 is absolutely reducible. Therefore, $reduct(C|P) = \{K_1, K_2, K_3, K_6\}$. According to Definition 16, we have $Pos_C(P) = Pos_{reduct(C|P)}(P) = \{x_1, x_3, x_4, x_5\}$, $BN_C(P) = BN_{reduct(C|P)}(P) = \{x_2\}$.

In Algorithm 1, relatively redundant elements are reduced at first, and absolutely redundant elements are reduced then. If the absolutely redundant elements are reduced firstly, K_3 and K_7 will be reduced in Example 2. Then, the originally relatively redundant elements, K_4 and K_5, will become relatively irreducible. In that way, we need one more rule with lower support degree to cover the objects x_3 and x_4. So, these two steps can not be exchanged.

6   Conclusions

The objective of knowledge reduction is to reduce redundant knowledge. Zhu proposed a reduction algorithm, but it can only reduce absolutely redundant knowledge. In order to mine maximal concise decision rules from covering approximation space, a relative reduction of covering approximation space is proposed in this paper, and an algorithm for knowledge reduction is developed. Its validity is illustrated by an example. It will be our future work to apply the method in knowledge acquisition and use it to solve some real life problems.

Acknowledgement

The authors would like to thank the anonymous referees for their valuable suggestions for improving this paper.

References

1. Bonikowski, Z., Bryniarski, E., Wybraniec, U.: Extensions and Intentions in the Rough Set Theory. Information Science 107, 149–167 (1998)
2. Hu, J., Wang, G.Y., Zhang, Q.H.: Uncertainty Measure of Covering Generated Rough Set. In: 2006 IEEE/WIC/ACM International Conference on Web Intelligence and Intelligent Agent Technology (WI-IAT 2006 Workshops) (WI-IATW 2006), pp. 498–504 (2006)
3. Huang, B., He, X., Zhou, X.Z.: Rough Entropy Based on Generalized Rough Sets Covering Reduction. Journal of Software 15(2), 215–220 (2004) (in Chinese)
4. Huang, H., Wang, G.Y., Wu, Y.: An approach for Incomplete Information Systems. In: Dasarathy, B.V. (ed.) Data Mining and Knowledge Discovery: Theory, Tools, and Technology VI. Proceedings of SPIE, vol. 5433, pp. 114–121 (2004)
5. Kryszkiewicz, M.: Rough Set Approach to Incomplete Information System. Information Sciences 112, 39–49 (1998)
6. Kryszkiewicz, M.: Rule in Incomplete Information Systems. Information Sciences 113, 271–292 (1999)
7. Li, T.J.: Rough Approximation Operators in Covering Approximation Spaces. In: Greco, S., Hata, Y., Hirano, S., Inuiguchi, M., Miyamoto, S., Nguyen, H.S., Słowiński, R. (eds.) RSCTC 2006. LNCS, vol. 4259, pp. 174–182. Springer, Heidelberg (2006)
8. Mordeson, J.N.: Rough Set Theory Applied to (fuzzy) Ideal Theory. Fuzzy Sets and Systems 121, 315–324 (2001)
9. Pal, S., Mitra, P.: Case generation using rough sets with fuzzy representation. IEEE Trans. On Knowledge and Data Engineering 16(3), 292–300 (2004)
10. Pawlak, Z.: Rough Sets. Computer and Information Science 11, 341–356 (1982)
11. Slowinski, R., Vanderpooten, D.: A Generalized Definition of Rough Approximations Based on Similarity. IEEE Tansactions on Knowledge and Data Engineering 12(2), 331–336 (2000)
12. Tsang, E.C.C., Chen, D., Lee, J.W.T., Yeung, D.S.: On The Upper Approximations of Covering Generalized Rough Sets. In: Proceedings of The Third International Conference on Machine Learning and Cybernetics, Shanghai, pp. 4200–4203 (2004)

13. Wang, G.Y.: Extension of Rough Set under Incomplete Information Systems. In: IEEE International Conference on Fuzzy Systems (FUZZ-IEEE), pp. 1098–1103 (2002)
14. Yang, Y., Wang, G.Y., Chen, P.J., Zhou, J., He, K.: Feature Selection in Audiovisual Emotion Recognition Based on Rough Set Theory. In: Peters, J.F., Skowron, A., Marek, V.W., Orłowska, E., Słowiński, R., Ziarko, W.P. (eds.) Transactions on Rough Sets VII. LNCS, vol. 4400, pp. 283–294. Springer, Heidelberg (2007)
15. Yuan, Z., Wu, Y., Wang, G.Y., Li, J.B.: Motion-Information-Based Video Retrieval System Using Rough Pre-classification. In: Peters, J.F., Skowron, A. (eds.) Transactions on Rough Sets V. LNCS, vol. 4100, pp. 306–333. Springer, Heidelberg (2006)
16. Zadeh, L.A.: Fuzzy Sets. Inform. Control 8, 338–353 (1965)
17. Zakowski, W.: Approximation in The Space (U, Π). Demonstratio Mathematica 16, 761–769 (1983)
18. Zhong, N., Yao, Y., Ohshima, M.: Peculiarity Oriented Multidatabase Mining. IEEE Trans. Knowledge and Data Eng. 15(4), 952–960 (2003)
19. Zhu, F.: On Covering Generalized Rough Sets, MS thesis, The University of Arizona, Tucson, Arizona, USA (2002)
20. Zhu, W., Wang, F.Y.: Reduction and Axiomization of Coving Generalized Rough Sets. Information Science 152, 217–230 (2003)
21. Zhu, W.: Topological Approaches to Covering Rough Sets. Information sciences 177, 1499–1508 (2007)
22. Zhu, W., Wang, F.Y.: On Three Types of Covering-based Rough Sets. IEEE transactions on knowledge and data engineering 19(8), 1131–1144 (2007)
23. Zhu, W., Wang, F.Y.: Covering Based Granular Computing for Conflict Analysis. In: Mehrotra, S., Zeng, D.D., Chen, H., Thuraisingham, B., Wang, F.-Y. (eds.) ISI 2006. LNCS, vol. 3975, pp. 566–571. Springer, Heidelberg (2006)
24. Zhu, W., Wang, F.Y.: A new type of covering rough sets. In: IEEE IS 2006, 4-6 September 2006, pp. 444–449 (2006)

Formal Description of the Cognitive Process of Memorization

Yingxu Wang

Theoretical and Empirical Software Engineering Research Centre (TESERC)
International Center for Cognitive Informatics (ICfCI)
Dept. of Electrical and Computer Engineering
Schulich School of Engineering, University of Calgary
2500 University Drive, NW, Calgary, Alberta, Canada T2N 1N4
Tel.: (403) 220 6141; Fax: (403) 282 6855
yingxu@ucalgary.ca

Abstract. Memorization is a key cognitive process of the brain because almost all human intelligence is functioning based on it. This paper presents a neuroinformatics theory of memory and a cognitive process of memorization. Cognitive informatics foundations and functional models of memory and memorization are explored toward a rigorous explanation of memorization. The cognitive process of memorization is studied that reveals how and when memory is created in long-term memory. On the basis of the formal memory and memorization models, the cognitive process of memorization is rigorously described using Real-Time Process Algebra (RTPA). This work is one of the fundamental enquiries on the mechanisms of the brain and natural intelligence according to the Layered Reference Model of the Brain (LRMB) developed in cognitive informatics.

Keywords: Cognitive informatics, cognitive computing, computational intelligence, neural informatics, brain science, memory, memorization, learning, knowledge representation, cognitive processes, memory creation, manipulation, modeling, LRMB, OAR.

1 Introduction

Memory as a faculty of information retention organs in the brain has been intensively studied in neural science, biopsychology, cognitive science, and cognitive informatics [1], [2], [5], [11], [13], [17], [30], [34]. However, memorization as a dynamic cognitive process that manipulates information among memories in the brain, particularly in the long-term memory has not been thoroughly investigated.

Definition 1. *Memory* is the physiological organs or networked neural clusters in the brain for retaining and retrieving information.

William James identified three components in human memory in 1890 known as *the after-image, the primary,* and *the secondary memory* [4]. The after-image memory is considered a relatively narrow concept because there are other sensorial inputs to the memory, such as hearing and touch. Thus, the after-image memory was gradually

M.L. Gavrilova et al. (Eds.): Trans. on Comput. Sci. V, LNCS 5540, pp. 81–98, 2009.

replaced by the concept of sensory memory. Contemporary theories on memory classification can be commonly described as the *sensory memory, short-term memory, and long-term memory* [1], [2], [8], [12], [13], [14], [15].

Examining the above types of memory it may be seen that there is a lack of an output-oriented memory, because the sensory memory is only an input-oriented buffer. The author and his colleagues introduce a new type of memory called the *action buffer memory* [34] that denotes the memory functions for the output-oriented actions, skills, and behaviors, such as a sequence of movement and a pre-prepared verbal sentence, which are interconnected with the motor servo muscles. Therefore, according to cognitive informatics, the logical architecture of memories in the brain can be classified into the following four categories: (a) the *sensory buffer memory*, (b) the *short-term memory*, (c) the *long-term memory*, and (d) the *action buffer memory*.

The contents of memory, particularly those in long-term memory, are information that may be classified into *knowledge, behavior, experience,* and *skills* [21, 23]. Therefore, the relationship between memory and knowledge is that of storage organs and contents. With the physiological basis of memories, memorization is a process of retention and retrieval about acquired information and past experience [15], [35], [36].

Definition 2. *Memorization* is a cognitive process of the brain at the meta- cognitive layer that establishes (encodes and retains) and reconstructs (retrieves and decodes) information in long-term memory.

This paper presents the cognitive informatics theory of memory and the cognitive process of memorization. Neural informatics foundations of memory and the relational model of memory are explored in Section 2. Logical models of memory, particularly the Object-Attribute-Relation (OAR) model, which form the context of human knowledge and intelligence, are explained in Section 3. The mechanisms of memorization as a cognitive process are investigated in Section 4, which explains how and when memory is created in long-term memory. On the basis of the memory and memorization models, the cognitive process of memorization is formally described using Real-Time Process Algebra (RTPA) in Section 5.

2   The Neural Informatics Foundations of Memory

Neural informatics [22], [34] is an interdisciplinary enquiry of the biological and physiological representation of information and knowledge in the brain at the neuron level and their denotational mathematical models [21], [26]. Neural informatics is a branch of cognitive informatics where memory is recognized as the foundation and platform of any natural or artificial intelligence.

2.1   Taxonomy of Memory

In neural informatics, the taxonomy of memory is categorized into four forms as given in the following cognitive model of memory.

Definition 3. The *Cognitive Model of Memory* (CMM) states that the logical architecture of human memory is parallel configured by the Sensory Buffer Memory (SBM), Short-Term Memory (STM), Conscious-Status Memory (CSM), Long-Term Memory (LTM), and Action-Buffer Memory (ABM), i.e.:

$$\text{CMM}\mathbf{ST} \triangleq \begin{array}{l} \text{SBM} \\ \| \ \text{STM} \\ \| \ \text{CSM} \\ \| \ \text{LTM} \\ \| \ \text{ABM} \end{array} \qquad (1)$$

where $\|$ denotes a parallel relations and $\mathbf{ST}$ represents an abstract system structural model.

The major organs that accommodate memories in the brain are the cerebrum or the cerebral cortex. In particular, the association and premotor cortex in the frontal lobe, the temporal lobe, sensory cortex in the frontal lobe, visual cortex in the occipital lobe, primary motor cortex in the frontal lobe, supplementary motor area in the frontal lobe, and procedural memory in cerebellum [36], [34]. The CMM model and the mapping of the four types of human memory onto the physiological organs in the brain reveal a set of fundamental mechanisms of neural informatics.

2.2 The Relational Metaphor of Memory

The conventional model of memory adopted in psychology is the *container* metaphor, which perceives that new information is stored in neurons of the brain. According to the container model, the brain needs an increasing number of neurons in order to store new information and knowledge acquired everyday. However, the observations in neural science and biopsychology indicates that the number of neurons of adult brains is relatively a constant at the level of about 10^{11} neurons [2], [7], [10] that will not increase during the entire life of a person.

Therefore, there is a need to seek a new model rather than the conventional container model to explain how information and knowledge are represented and retained in the brain. For this purpose, a relational model of human memory is developed as described below.

Definition 4. The *relational model of memory* is a logical memory model that states information is represented and retained in the memory by relations, which is embodied by the synaptic connections among neurons.

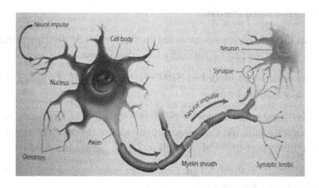

Fig. 1. The micro model of memory (Sternberg, 1998 [15])

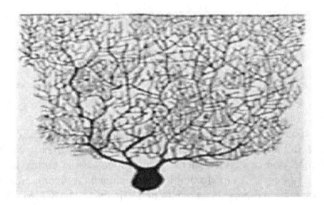

Fig. 2. The macro model of memory (Sternberg, 1998 [15])

The *relational* metaphor indicates that the brain does not create new neurons to represent newly acquired information; instead, it generates new synapses between the existing neurons in order to represent new information.

The *micro* and *macro* models of memory, particularly LTM, can be illustrated in Figs. 1 and 2, respectively, which are supported by observations in neuroscience and neuropsychology [2], [7], [15].

Theorem 1. Properties of LTM are as follows:

- It is dynamic;
- It is directed, i.e. relations $r(\alpha, \beta) \neq r(\beta, \alpha)$ where α and β are two different neurons $\alpha \neq \beta$;
- It is reconfigurable;
- It can be strengthened by frequently accesses;
- It contains loops;
- It can be traversed or searched;
- It cannot be sorted.

2.3   Functional Models of Memory

Corresponding to the CMM model as given in Definition 3, the functional models of the four types of memories can be formally modeled below.

Definition 5. The functional model of *SBM* is a set of *parallel queues* corresponding to each of the sensors of the brain.

Definition 6. The functional model of *STM* is a set of *temporal and plastic neural clusters* that accommodates the thinking threads in the form of relations and links to related objects in other part of STM, as well as LTM, SBM, and ABM.

Definition 7. The functional model of *LTM* is *hierarchical neural clusters* with partially connected *neurons* via *synapses*.

Definition 8. The functional model of *ABM* is a set of *parallel queues*, where each of them represents a sequence of actions or a process.

Definition 9. The functional model of *CSM* is a combination of the forms of LTM and STM, where the persistent statuses of the brain and body are maintained in LTM, while the interactive and real-time statuses are retained in STM before they are updated into the LTM form of CSM.

The reconfigurable neural clusters of STM cohere and connect related objects such as images, data, and concepts, and their attributes by synapses in order to form contexts and threads of thinking. Therefore, the main function of STM may be analogized to an index memory connecting to other memories, particularly LTM.

STM is the working memory of the brain. The capacity of STM is much smaller than that of LTM, but it is hundred times greater than 7 ± 2 digits as Miller proposed [9]. Limited by the temporal space of STM, one has to write complicated things on paper or other types of external memories in order to compensate the required working memory space in a thinking process.

Theorem 2. The *dynamic neural cluster model* states that LTM is dynamic. New neurons (to represent objects or attributes) are assigning, and new connections (to represent relations) are creating and reconfiguring all the time in the brain.

3   The Logical Model of Memory

The neural informatics model of memory has been developed in Section 2. This section describes the logical model of memory by investigating the form of knowledge representation in the brain. Based on the logical models of memory, the capacity of human memory may be formally estimated and mechanisms of the memorization process may be rigorously explained.

3.1   The OAR Model of Memory

To rigorously explain the hierarchical and dynamic neural cluster model of memory at physiological level, a logical model of memory is needed as given below known as the Object-Attribute-Relation (OAR) model.

Definition 10. The *OAR model* of LTM can be described as a triple, i.e.:

$$OAR \triangleq (O, A, R) \tag{2}$$

where O is a finite nonempty set of objects identified by unique symbolic names, i.e.:

$$O = \{o_1, o_2, ..., o_i, ..., o_n\} \tag{3}$$

For each given $o_i \in O$, $1 \le i \le n$, A_i is a finite nonempty set of attributes for characterizing the object o_i, i.e.:

$$A_i = \{A_{i1}, A_{i2}, ..., A_{ij}, ..., A_{im}\} \tag{4}$$

where each $o_i \in O$ or $A_{ij} \in A_i$, $1 \le i \le n$, $1 \le j \le m$, is physiologically implemented by a neuron in the brain.

For each given $o_i \in O$, $1 \le i \le n$, R_i is a finite nonempty set of relations between o_i and other objects or attributes of other objects, i.e.:

$$R_i = \{R_{i1}, R_{i2}, ..., R_{ik}, ..., R_{iq}\} \tag{5}$$

where R_{ik} is a relation between two objects, o_i and $o_{i'}$, and their attributes A_{ij} and $A_{i'j}$, $1 \leq i \leq n$, $1 \leq j \leq m$, i.e.:

$$
\begin{aligned}
R_{ik} = \ & r\,(o_i, o_{i'}) \\
& | \ r\,(o_i, A_{ij}) \\
& | \ r\,(A_{ij}, o_{i'}) \\
& r\,(A_{ij}, A_{i'j}), \ 1 \leq k \leq q
\end{aligned}
\tag{6}
$$

To a certain extent, the entire knowledge in the brain can be modeled as a global OAR model as given in Fig. 3.

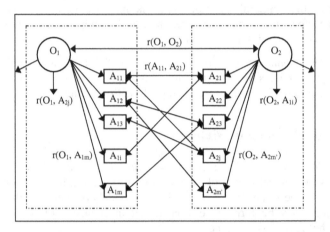

Fig. 3. The OAR model of logical memory architectures

3.2   The Extended OAR Model of Memory

The OAR model developed in the preceding subsection reveals a generic abstract model of LTM and the form of internal representation of learning and other cognitive activities known as knowledge, behavior, experience, and skills. Mapping it onto the cognitive structure of the brain, an extended OAR model of the brain, EOAR, is given in Fig. 4, where the external world is represented by *real entities* (RE), and the internal world by *virtual entities* (VE) and *objects* (O). The internal world can be divided into two layers: the *image* layer and the *abstract* layer.

Definition 11. The *extended OAR model* of the brain, *EOAR*, states that the external world is represented by *real entities*, and the internal world by *virtual entities* and *objects*. The internal world can be divided into two layers known as the *image layer* and the *abstract layer*.

The virtual entities are direct images of the external real-entities located at the image layer. The objects are abstract artifacts located at the abstract layer. The abstract layer is an advanced property of human brains. It is noteworthy that animal species have no such abstract layer in their brains in order to support *abstract* or *indirect* thinking and reasoning [34]. In other words, high-level abstract thinking is a unique power of the human brain known as the *qualitative* advantage of human brains. The other advantage of the human brain is the tremendous capacity of LTM in the cerebral cortex known as the *quantitative* advantages. On the basis of these two principal advantages, mankind gains the power as human beings.

There are *meta-objects* (O) and *derived objects* (O') at the abstract layer. The former are concrete objects directly corresponding to the virtual entities and then to the external world. The latter are abstracted objects that are derived internally and have no direct connection with the virtual entities or images of the real-entities such as abstract concepts, notions, numbers, and artifacts. The objects on the brain's abstract layer can be further extended into a network of objects, attributes, and relations according to the EOAR model as shown in Fig. 4. The connections between objects/attributes (O/A) via relations are *partially* connected rather than fully connected. In other words, it is not necessary to find a relation among all pairs of objects or attributes.

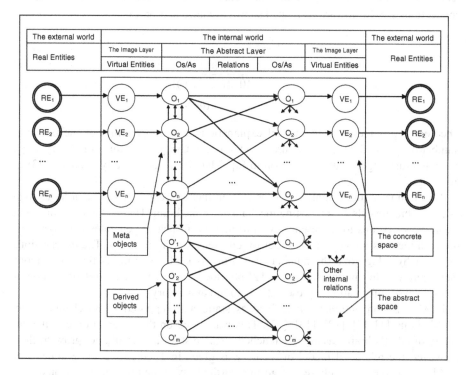

Fig. 4. The EOAR model of the brain

It is noteworthy that the higher level cognitive processes and consciousness, such as willingness, emotions, desires, and attributes are results of both such internal states in the brain and current external stimuli. Detailed discussions may be referred to the LRMB model [35]. It is also noteworthy that the cognitive model of the brain is looped. This means that an internal virtual entity is not only abstracted from the real-entity as shown on the left-hand side in Fig. 4, but also eventually connected to the entities on the right-hand side. This is the foundation of thinking, reasoning, learning, and other high-level cognitive processes, in which internal information has to be related to the real-world entities, in order to enable the mental processes meaningfully embodied to real-world semantics.

3.3 The Capacity of Human Memory

It is observed that the total neurons in the brain is about $n = 10^{11}$ and their *average* synaptic connections is $s = 10^3$ [2], [7], [10]. According to the relational model of memory, the fundamental question on the capacity of human memory derived in cognitive science and neuropsychology can be reduced to a classical combinatorial problem [33].

Theorem 3. The capacity of human memory C_m is determined by the total potential relational combinations, C_n^s, among all neurons $n = 10^{11}$ and their *average* synaptic connections $s = 10^3$ to various related subset of entire neurons, i.e.:

$$C_m \triangleq \mathbf{C}_n^s$$
$$= \frac{10^{11}!}{10^3!(10^{11}-10^3)!} \quad (7)$$
$$= 10^{8,432} \quad [bit]$$

Theorem 3 provides a mathematical explanation of the upper limit of the potential number of connections among neurons in the brain. Using approximation theory and a computational algorithm, the solution to Eq. 7 had been successfully obtained [33] as given above.

The finding on the magnitude of the human memory capacity on the order as high as $10^{8,432}$ bits reveals an interesting mechanism of the brain. That is, the brain does not create new neurons to represent new information, instead it generates new synapses between the existing neurons in order to represent new information. The observations in neurophysiology that the number of neurons is kept stable rather than continuous increasing in adult brains [7], [10], [12] provided evidences for the relational cognitive model of information representation in human memory.

LTM was conventionally perceived as static and supposed to no change in an adult's brain [1], [4], [12], [13], [15]. This was based on the observation that the capacity of adult's brain has already reached a stable stage and would not grow further. However, the relational model of memory as given in Theorems 2 and 3 states that LTM is dynamic and lively reconfiguring, particularly at the lower levels or on leaves of the neural clusters. Otherwise, one cannot explain the mechanism of memory establishment and update [12], [14], [34].

Actually, the two perceptions above are not contradictory. The former observes that the macro-number of neurons will not change significantly in an adult brain. The latter reveals that information and knowledge are physically and physiologically retained in LTM via newly created synapses between neurons rather than the neurons themselves.

4   Mechanisms of Memorization

On the basis of formal models of memory at the physiological and logical levels as developed in Sections 2 and 3, this section attempts to rigorously explores the mechanisms of memorization and its cognitive process.

4.1   Memorization as a Cognitive Process

According to Definition 2, the process of memorization encompasses *encoding* (knowledge representation), *retention* (store in LTM), *retrieve* (LTM search), and *decoding* (knowledge reformation) as shown in Fig. 5. The sign of a successful memorization process in cognitive informatics is that the same information can be correctly recalled or retrieved. Therefore, memorization may need to be repeated or rehearsed for a number of cycles before it is completed.

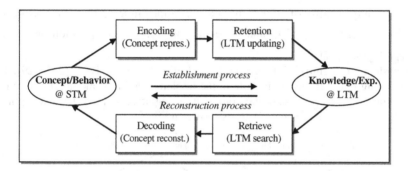

Fig. 5. The process of memorization

It is noteworthy that the memorization process is a closed-loop between STM and LTM, where it may be divided into the establishment and reconstruction phases.

Definition 12. The *establishment phase* of memorization is a memory creation process that represents a certain information in the form of a sub-OAR in STM via encoding, and then creates relations with the entire OAR in LTM via retention.

Definition 13. The *reconstruction phase* of memorization is a retrieval process that searches the entire OAR in LTM via content patterns or keywords, and then reconfigures the information in STM via decoding.

It is recognized that computers store data in a direct and unconsumed manner; while the brain stores information by relational neural clusters. The former can be accessed directly by explicit addresses and can be sorted; while the latter may only be retrieved by content-sensitive search and matching among neuron clusters where spatial connections and configurations themselves represent information. The tremendous difference of memory magnitudes between human beings and computers demonstrates the efficiency of information representation, storage, and processing in human brains.

4.2 How Memory Is Created?

As learning is aimed at acquiring new knowledge based on comprehension [32], memorization is required to create or update LTM by searching and analyzing the contents of STM and selecting useful (i.e. most frequently used) information into LTM.

According to the *OAR* model, the result of knowledge acquisition or learning can be embodied by the updating of the existing *OAR* in the brain.

Theorem 4. The entire knowledge model maintained in the brain states that the internal memory or the representation of learning results in the form of the OAR structure, which can be updated by concept compositions ⊎ between the existing OAR and the newly created sub-OAR (sOAR), i.e.:

$$OAR'\textbf{ST} \triangleq OAR\textbf{ST} \uplus sOAR\,\textbf{ST}$$

$$= OAR\textbf{ST} \uplus (O_s, A_s, R_s) \tag{8}$$

where **ST** is a type suffix of system structure as defined in *Real-Time Process Algebra* (RTPA) [18], [21], [24], [29], and ⊎ denotes the concept composition operation in *Concept Algebra* [27].

According to cognitive informatics [17], [19], [20], [21], [22], [25], [29], [30], [31], sleeping plays an important role in the implementation of memorization. Sleeping is a subconscious process of the brain that its cognitive and psychological purpose is to update LTM in the form of OAR as shown in Fig. 6.

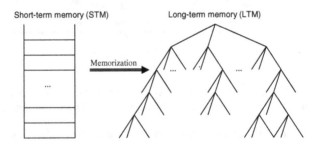

Fig. 6. Memorization as information transforming from STM to LTM

Theorem 5. The *mechanisms of memorization* characterized by OAR updating in LTM is based on the following selective criteria:

- A sub-OAR in STM was frequently or repetitively used in the previous 24 hours;
- A sub-OAR was related to the entire OAR in LTM at a higher level of the neural cluster hierarchy;
- A sub-OAR was given special attention or a higher retention weight.

Corollary 1. The *algorithm of memorization* can be described by the following steps:

- To identify association to existing knowledge structure in the LTM clusters in the form of OAR;

- To generate physiological neural links between new and existing objects by synapses, if there are existing or related knowledge clusters in LTM;
- To create a new sub-OAR cluster, if there is no existing or related knowledge cluster in LTM.

It can be seen that the third step stated in Corollary 1 is the hardest case in memorization. Based on the memorization algorithm, the relationship between learning and memorization becomes apparent. The former is a front-end process to acquire and represent knowledge in the form of sub-OARs; while the latter is a back-end process to create or update the OAR model of entire knowledge in LTM by knowledge composition as defined in Eq. 8.

Typical *memory devices* identified are categorization, organization, clustering, acronym/acrostics, interactive imagery, visualization, highlight keywords, rehearsal, and elaboration [3], [11]. It may be observed that *rehearsal* and *repetitive processing* of the same information play a crucial role in memorization.

Corollary 2. The longer the time spent on memorization and learning, the better the effect of memorization.

Corollary 2 indicates that time of concentration is the only magic in learning, memorization, and knowledge creation. Therefore, the fundamental approach to improve memorization and creative research is both concentration and sufficient time investment.

4.3 When Memory Is Created?

The cognitive model of the brain [34] classifies life functions of the brain into two categories known as the conscious and subconscious ones. The latter are inherited subconscious processes and cannot be intentionally changed; while the former are acquired and can be programmed consciously by certain motivations and goals. It is noteworthy that the subconscious life functions play an important role in parallel with the conscious counterparts. That is, the higher layer cognitive processes are implemented based on the support of the underlying subconscious ones at the lower layers according to the LRMB model [35]. Therefore, a study on the subconscious behaviors of the brain and their mechanisms may be the key to understand how the brain works.

The investigation on the subconscious aspect of memorization may be focused on the following questions: a) When is the memorization process completed in LTM? b) Why do all mammals need sleep? and c) What is the cognitive mechanism of sleep?

Sleep as an important physiological and psychological phenomenon was perceived as innate, and few hypotheses and theories have been developed to explain the reason [6], [16]. The following theories explain the roles of sleep in LTM establishment.

Lemma 1. The memory in LTM is established during sleeping.

Lemma 1 is supported by the following observations and experiments. A group of UK scientists observed that stewardesses serving long-haul flights had bad memory in common [21]. An explanation about the reason of this phenomenon was that the stewardesses have been crossing time zones too frequently! However, according to Lemma 1, the memory problems of stewardesses were caused by the lack of quality sleep during night flights. As a consequence, the LTM could not be properly built.

Lemma 1 logically explains the following common phenomena: (a) All mammals, including human beings, need to sleep; (b) When sleeping, the blood supply to the brain reaches the peak, at about 1/3 of the total consumption of the entire body. However, during daytime the brain just consumes 1/5 of the total blood supply in the body [6], [15], [36]; and (c) According to the cognitive model of the brain [34], human beings are naturally an intelligent real-time information processing system. Since the brain is busy during day-time, it is logical to schedule the functions of LTM establishment at night, when more processing time is available and fewer inference or interruptions occur due to external events.

Based on Lemma 1, the following cognitive informatics *theory of sleeping* can be derived.

Theorem 6. *Long-term memory establishment* is a subconscious process that its major mechanism is by sleeping, i.e.:

$$\text{Cognitive purpose of sleep} = \text{LTM establishment} \qquad (9)$$

Theorem 6 describes an important finding on one of the fundamental mechanisms of the brain and the cognitive informatics meaning of sleep, although there are other physiological purposes of sleep as well, such as resting the body, avoid dangers, and saving energy.

Corollary 3. Lack of sleep results in bad memory, because the memory in LTM cannot be properly established.

Corollary 4. The subconscious cognitive processes of the brain do not sleep throughout the entire human life.

It was commonly believed that heart is the only organ in human body that never takes rest during the entire life. However, Corollary 4 reveals that so does the brain. The non-resting brain is even more important than heart because the latter is subconsciously controlled and maintained by the former.

Based on Lemma 1 and Theorem 5, the following principle on memorization can be established.

Theorem 7. The *24-hour law of memorization* states that the general establishment cycle of LTM is equal to or longer than 24 hours, i.e.:

$$LTM\ establishment\ cycle \geq 24\ \text{[hrs]} \qquad (10)$$

where the 24-hour cycle includes any kind of combinations of awake, asleep, and siesta.

5   Formal Description of the Memorization Process

The physiological and neural informatics foundation of memorization is the dynamic updating of the LTM in the logic form of the OAR model. This section presents a formal treatment of memorization as a cognitive process. Based on the cognitive process perception, a formal algorithm and a rigorous RTPA model for explaining the memorization process are developed.

5.1 The Memorization Process and Algorithm

As illustrated in Fig. 5, memorization as a cognitive process can be described by two-phases: the establishment phase and the reconstruction phase. The former represents the target information in the form of OAR and creates the memory in LTM. The letter retrieves the memorized information and reconstructs it in the form of a concept in STM. Memorization can also be perceived as the transformation of information and knowledge between STM and LTM, where the forward transformation from STM to LTM is for memory establishment, and the backward transformation from LTM to STM is for memory reconstruction.

The logical model of the memorization process can be described as shown in Fig. 7. Based on Fig. 7, a memorization algorithm is elaborated as follows.

Algorithm 1. The cognitive process of memorization can be carried out by the following steps:

- (0) *Begin*
- (1) *Encoding*: This step generates a representation of a given concept by transferring it into a sub-OAR model;
- (2) *Retention*: This step updates the entire OAR in LTM with the sub-OAR for memorization by creating new synaptic connections between the sub-OAR and the entire OAR;
- (3) *Rehearsal test:* This step checks if the memorization result in LTM needs to be rehearsed. If yes, it continues to practice Steps (4) and (5); otherwise, it jumps to Step (7);
- (4) *Retrieval:* This step retrieves the memorized object in the form of sub-OAR by searching the entire OAR with clues of the initial concept;
- (5) *Decoding:* This step transfers the retrieved sub-OAR from LTM into a concept and represents it in STM;
- (6) *Repetitive memory test:* This step tests if the memorization process was succeeded or not by comparing the recovered concept with the original concept. If need, repetitive memorization will be called.
- (7) *End.*

It is noteworthy that the input of memorization is a structured concept formulated by learning.

5.2 Formal Description of the Memorization Process

The cognitive process of memorization described in Algorithm 1 and Fig. 7 can be formally modeled using RTPA [18], [21], [28] as given in Fig. 8. According to the LRMB model [Wang et al., 2006] and the OAR model [23] of internal knowledge representation in the brain, the input of the memorization process is a structured concept $c(O\mathbf{S}, A\mathbf{S}, R\mathbf{S})\mathbf{ST}$, which will be transformed to update the entire OAR model of knowledge in LTM in order to create a permanent memory. Therefore, the output of memorization is the updated $OAR'\mathbf{ST}$ in LTM.

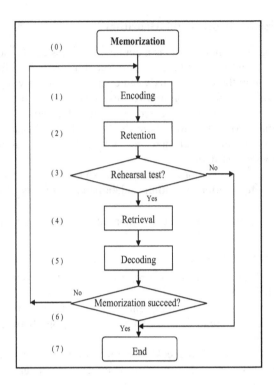

Fig. 7. The cognitive process of memorization

In the formal memorization process as shown in Fig. 8, the *encoding* subprocess is modeled as a function that maps the given concept *c***ST** into a *sOAR***ST**. The *retention* subprocess composes the *sOAR***ST** with the entire *OAR***ST** in LTM that maintains the whole knowledge of an individual. In order to check the memorization quality, rehearsals may usually be needed. In a rehearsal, the *retrieval* subproecss searches a related *sOAR***ST** in LTM by giving clues of previously memorized objects and attributes in *c***ST**. Then, the *decoding* subprocess transfers the *sOAR***ST** into a recovered concept *c'***ST**. In the repetitive memory test subprocess, the reconstructed *c'***ST** will be compared with the original input of *c***ST** in order to determine if further memorization is recursively needed.

According to the 24-hour law of memorization as stated in Theorem 7, the memorization process may be completed with a period longer than 24 hours by several cycles of repetitions. Although, almost all steps in the process as shown in Fig. 7 are conscious, the key step of *retention* is subconscious or non intentionally controllable. The rules of thumb of high quality retention have been described in Theorem 5.

Based on the LRMB model [35], the memorization process is closely related to learning [24]. In other words, memorization is a back-end process of learning, which retains learning results in LTM and retrieves them when rehearsals are needed. The retrieve process is search-based by concept or sOAR matching.

```
┌─────────────────────────────────────────────────────────┐
│              The Memorization Process                    │
│                                                          │
│ Memorization (I:: c(OS, AS, RS)ST; O:: OAR'ST) ≙         │
│ {I. Encoding                                             │
│ c(OS, AS, RS)ST → sOARST                                 │
│ // Concept representation                                │
│                                                          │
│ II. Retention                                            │
│ → OAR'ST := OARST ⊎ sOARST                               │
│ // Update OARST in LTM                                   │
│                                                          │
│ III. Rehearsal                                           │
│ → ◆ Rehearsal BL = T                                     │
│     (IV. Retrieval                                       │
│        ↣ Search (I:: OARST;                              │
│                                                          │
│                   O:: sOARST | (OS, AS, RS)ST⊆ OARST))   │
│        // Retrieval sOARST in LTM                        │
│                                                          │
│     V. Decoding                                          │
│        → (sOARST →  c'(OS, AS, RS)ST)                    │
│        // Concept reconstruction                         │
│     )                                                    │
│ VI. Repeat                                               │
│ → ◆ (c'(OS, AS, RS)ST) ~ c(OS, AS, RS)ST)                │
│        → ⊗ // Memorization succeed                       │
│     | ◆ ~                                                │
│        → Memorization (I:: c(OS, AS, RS)ST; O:: OAR'ST)  │
│        // Retry                                          │
│ }                                                        │
└─────────────────────────────────────────────────────────┘
```

Fig. 8. Formal description of the memorization process in RTPA

It is noteworthy that the memorization process is a fully creative process, which generates new sub-OARs and establishes physiological representations of them with the existing OAR in LTM by new synaptic connections. Therefore, in some extent, memorization is a subconscious physiological process where new synapses have to be grown inside the brain over time in order to transfer learnt information or knowledge into permanent memory.

6   Conclusions

This paper has presented a theory of memory and the cognitive process of memorization. Memorization has been identified as a key cognitive process of the brain because almost all human intelligence is functioning based on it. Neural informatics foundations and function models of memory and memorization have been explored in this paper. Logical models of memory, particularly the Object-Attribute-Relation (OAR) model have been developed, which form the context of human knowledge and intelligence.

Some enlightening findings on memory and memorization in cognitive informatics are as follows:

- LTM establishment is a subconscious process;
- The LTM is established during sleep;
- The major mechanism for LTM establishment is by sleeping;
- The general acquisition cycle of LTM equals to or is longer than 24 hours;
- The mechanism of LTM establishment is to update the entire memory of information represented as an OAR model in the brain;
- Eye movement and dreams play an important role as the observable indicator in LTM creation.

The mechanisms of memorization have been rigorously explored as a cognitive process, and the fundamental queries on how and when memory is created in long-term memory have been logically explained.

Acknowledgements

The author would like to acknowledge the Natural Science and Engineering Council of Canada (NSERC) for its partial support to this work. The author would like to thank the anonymous reviewers for their valuable comments and suggestions.

References

1. Baddeley, A.: Human Memory: Theory and Practice. Allyn and Bacon, Needham Heights (1990)
2. Gabrieli, J.D.E.: Cognitive Neuroscience of Human Memory. Annual Review of Psychology 49, 87–115 (1998)
3. Gray, P.: Psychology, 2nd edn. Worth Publishers, Inc., New York (1994)
4. James, W.: Principles of Psychology, New York, Holt (1890)
5. Leahey, T.H.: A History of Psychology: Main Currents in Psychological Thought, 4th edn. Prentice-Hall Inc., Upper Saddle River (1997)
6. Maquet, P.: The Role of Sleep in Learning and Memory. Science 294(5544), 1048–1051 (2001)
7. Marieb, E.N.: Human Anatomy and Physiology, 2nd edn. The Benjamin/Cummings Publishing Co., Inc., Redwood City (1992)
8. Matlin, M.W.: Cognition, 4th edn. Harcourt Brace College Publishers, Orlando (1998)
9. Miller, G.A.: The Magical Number Seven, Plus or Minus Two: Some Limits of our Capacity for Processing Information. Psychological Review 63, 81–97 (1956)
10. Pinel, J.P.J.: Biopsychology, 3rd edn. Allyn and Bacon, Needham Heights (1997)
11. Reisberg, D.: Cognition: Exploring the Science of the Mind, 2nd edn. W.W. Norton & Company, Inc. (2001)
12. Rosenzmeig, M.R., Leiman, A.L., Breedlove, S.M.: Biological Psychology: An Introduction to Behavioral, Cognitive, and Clinical Neuroscience, 2nd edn. Sinauer Associates, Inc., Publishers, Sunderlans (1999)
13. Smith, R.E.: Psychology. West Publishing Co., St. Paul (1993)
14. Squire, L.R., Knowlton, B., Musen, G.: The Structure and Organization of Memory. Annual Review of Psychology 44, 453–459 (1993)

15. Sternberg, R.J.: In Search of the Human Mind, 2nd edn. Harcourt Brace & Co., Orlando (1998)
16. Stickgold, R., Hobson, J.A., Fosse, R., Fosse, M.: Sleep, Learning, and Dreams: Off-Line Memory Reprocessing. Science 294(5544), 1048–1051 (2001)
17. Wang, Y.: Keynote: On Cognitive Informatics. In: Proc. 1st IEEE International Conference on Cognitive Informatics (ICCI 2002), Calgary, Canada, pp. 34–42. IEEE CS Press, Los Alamitos (2002a))
18. Wang, Y.: The Real-Time Process Algebra (RTPA). Annals of Software Engineering: An International Journal 14, 235–274 (2002b)
19. Wang, Y.: On Cognitive Informatics, Brain and Mind: A Transdisciplinary. Journal of Neuroscience and Neurophilosophy 4(3), 151–167 (2003)
20. Wang, Y.: Keynote: Cognitive Informatics - Towards the Future Generation Computers that Think and Feel. In: Proc. 5th IEEE International Conference on Cognitive Informatics (ICCI 2006), Beijing, China, pp. 3–7. IEEE CS Press, Los Alamitos (2006)
21. Wang, Y.: Software Engineering Foundations: A Software Science Perspective. CRC Book Series in Software Engineering, vol. II. Auerbach Publications, NY (2007a)
22. Wang, Y.: The Theoretical Framework of Cognitive Informatics. International Journal of Cognitive Informatics and Natural Intelligence 1(1), 1–27 (2007b)
23. Wang, Y.: The OAR Model of Neural Informatics for Internal Knowledge Representation in the Brain. International Journal of Cognitive Informatics and Natural Intelligence 1(3), 64–75 (2007c)
24. Wang, Y.: The Theoretical Framework and Cognitive Process of Learning. In: Proc. 6th International Conference on Cognitive Informatics (ICCI 2007). IEEE CS Press, CA (2007d)
25. Wang, Y.: Keynote, On Abstract Intelligence and Its Denotational Mathematics Foundations. In: Proc. 7th IEEE International Conference on Cognitive Informatics (ICCI 2008), Stanford University, CA, USA, pp. 3–13. IEEE CS Press, Los Alamitos (2008a)
26. Wang, Y.: On Contemporary Denotational Mathematics for Computational Intelligence. Transactions of Computational Science 2, 6–29 (2008b)
27. Wang, Y.: On Concept Algebra: A Denotational Mathematical Structure for Knowledge and Software Modeling. International Journal of Cognitive Informatics and Natural Intelligence 2(2), 1–19 (2008c)
28. Wang, Y.: RTPA: A Denotational Mathematics for Manipulating Intelligent and Computational Behaviors. International Journal of Cognitive Informatics and Natural Intelligence 2(2), 44–62 (2008d)
29. Wang, Y.: On Abstract Intelligence: Toward a Unified Theory of Natural, Artificial, Machinable, and Computational Intelligence. International Journal of Software Science and Computational Intelligence 1(1), 1–17 (2009a)
30. Wang, Y., Kinsner, W., Anderson, J.A., Zhang, D., Yao, Y.Y., Sheu, P., Tsai, J., Pedrycz, W., Latombe, J.-C., Zadeh, L.A., Patel, D., Chan, C.: A Doctrine of Cognitive Informatics. Fundamenta Informaticae 90(3), 1–26 (2009a)
31. Wang, Y., Zadeh, L.A., Yao, Y.: On the System Algebra Foundations for Granular Computing. International Journal of Software Science and Computational Intelligence 1(1), 64–86 (2009b)
32. Wang, Y., Davrondjon, G.: The Cognitive Process of Comprehension. In: Proc. 2nd International Conference on Cognitive Informatics (ICCI 2003), pp. 93–97. IEEE CS Press, London (2003)

33. Wang, Y., Liu, D., Wang, Y.: Discovering the Capacity of Human Memory, Brain and Mind: A Transdisciplinary. Journal of Neuroscience and Neurophilosophy 4(2), 189–198 (2003)
34. Wang, Y., Wang, Y.: Cognitive Informatics Models of the Brain. IEEE Transactions on Systems, Man, and Cybernetics (C) 36(2), 203–207 (2006)
35. Wang, Y., Wang, Y., Patel, S., Patel, D.: A Layered Reference Model of the Brain (LRMB). IEEE Transactions on Systems, Man, and Cybernetics (C) 36(2), 124–133 (2006)
36. Wilson, R.A., Keil, F.C.: The MIT Encyclopedia of the Cognitive Sciences. MIT Press, Cambridge (2001)

Intelligent Processing of an Unrestricted Text in First Order String Calculus

Andrew Gleibman

Sampletalk Technologies, POB 7141, Yokneam-Illit 20692, Israel
gleibman@sampletalk.com

Abstract. First Order String Calculus (FOSC), introduced in this paper, is a generalization of First Order Predicate Calculus (FOPC). The generalization step consists in treating the unrestricted strings, which may contain variable symbols and a nesting structure, similarly to the predicate symbols in FOPC. As a logic programming technology, FOSC, combined with a string unification algorithm and the resolution principle, eliminates the need to invent logical atoms. An important aspect of the technology is the possibility to apply a matching of the text patterns immediately in logical reasoning. In this way the semantics of a text can be defined by string examples, which only demonstrate the concepts, rather than by a previously formalized mathematical knowledge. The advantages of avoiding this previous formalization are demonstrated. We investigate the knowledge representation aspects, the algorithmic properties, the brain simulation aspects, and the application aspects of FOSC theories in comparison with those of FOPC theories. FOSC is applied as a formal basis of logic programming language Sampletalk, introduced in our earlier publications.

1 Introduction

In this work we address one of the most intriguing questions of cognitive informatics: *What kind of knowledge can be extracted from something essentially new, which is not relevant to anything known at the moment?* Simulating such cognition, we are trying to simulate the formation of substantially new knowledge in our brain. The only background knowledge, which we assume in our consideration, is the knowledge that a text has a sequential structure. We apply this assumption simultaneously for the perceived objects (that is, we are interested in perceiving and understanding the unrestricted texts) and for the means to simulate the analysis of such objects in the brain. The existing cognition and machine learning techniques usually assume less trivial frameworks. For example, the classification algorithms typically are based on geometrical properties of a feature space, which, in its turn, depends on the algorithms for extracting numerical feature values from the perceived objects.

So, we limit ourselves in a text environment and try to find the most universal features of this environment, which are independent of any particular knowledge about the text structure, the world described in the text, the language the text is written in, and the status and intentions of the writer. Still more important, we try to be independent, as much as possible, of any formal models of known phenomena. We will

M.L. Gavrilova et al. (Eds.): Trans. on Comput. Sci. V, LNCS 5540, pp. 99–127, 2009.

see that this *minimalist* approach not only leads to a simulation of the processes, occurring in the human brain, but provides a rigorous framework for building practical intelligent text processing systems.

Let us start with an analysis of the role some *artificial* symbolic notations, such as logical predicates, play in the attempts to describe and understand the external world. These notations are usually applied for expressing *known* relations between *known* objects in the context of other *known* objects, terms, relations and notations. Note that the choice of the notations and relations reflects the status of the observer rather than the inherent features of the perceived objects. This observation is crucially important for us. The cognition, based on artificial notations, assumes the usage of some previous knowledge, which may be relevant or irrelevant to the perceived world. The more sophisticated is our set of notations, the more dependent is our cognition on our previous knowledge. We will discuss some negative aspects of this knowledge and look for a way to avoid them.

A symbolic representation, which unambiguously denotes complex objects and relations, has been regarded as the main attribute of any mathematical formalization. Alonzo Church [3] describes the notations for expressing senses, or meanings, as *complex names* and assumes the following rule for manipulating them: *the sense of a complex name is not changed if we replace one of its components with another complex name, whose sense is the same as that of the replaced component.* This assumption serves as a basis for most attempts to make a flexible formal model of the external world. In the context of predicate logic the *complex names* are composed from logical terms, predicates, connectives and other symbols.

Let us look at this assumption from a specific point of view: the complexity and elaboration of the *artificially* created complex names tends to grow, making them more and more difficult for composition, combination and comprehending by a human. Over centuries such complex names are abundant with Greek letters, indices, brackets, superscripts, complex syntax, vernacular designations etc. Usually every object and relation, subjected to formalization or modeling on a computer, is notated with a specially devised artificial symbolic notation. The notations often need to be accompanied with informal comments or substantial documents, which help understand the meaning of the involved symbols.

Do we always need so elaborated *artificial* symbolic notation to abide to Church's assumption mentioned above? Can we apply, instead of the artificial notations, the notations which already exist in *natural* sources, such as unrestricted natural language texts or biological sequences? Do we always need to apply a *prior* semantic knowledge, expressed mathematically, in order to make inferences from such data?

Interestingly, we can formalize our knowledge and define algorithms without complex artificial syntax at all. Known text examples, such as encyclopedia articles and patent formulations, show that we can describe very complex things without explicitly defined formal notations and appeal only to the common meaning of natural language words and expressions. Can we do this in a formal language? Can we apply *natural* phrases and their senses in algorithm definition? Can we use natural language words and concepts in a computer program as fluently as in human speech, without having to *explicitly* define the artificial *complex names* for that? Can we extract and apply patterns of unrestricted texts immediately in reasoning, consistently with the meaning of these texts? Is there a possibility to define what this meaning is without reference to any mathematical constructions known a priori?

We will clarify these questions by challenging the concept of logical predicate. For this purpose we develop a special kind of logic, which can be termed *a predicateless logic* or *a string calculus*. Using this logic, we can compose a *formal* theory or an algorithm using *natural* words and phrases that do not have explicit formal definitions.

On the first glance this seems impossible. Note, however, that in our speech we can correctly apply numerous words without knowing the *explicit* definition of their senses. We met these words in the phrases, which we understand, and we are able to apply these words in a similar context. We have learned them *implicitly*, via the contexts. Can we do this on a computer? Can we apply unrestricted words, which are not defined explicitly, in a formal definition that a computer may understand, consistently with the common meaning of these words? If we can do so, we can *immediately* employ the meaning of numerous natural language words and phrases, existing in such knowledge sources as work instructions, encyclopedia articles and patent formulations, in order to make inferences from this knowledge.

This work should be considered an attempt to do this in a Logic Programming framework. Logic Programming systems typically are based on the following two fundamental principles: *a logic term unification principle* and *a resolution principle*. The first principle allows manipulation with senses, according to Church's assumption mentioned above, by unification of logical terms: *term constituents, placed in similar positions of the unified terms, have similar senses*. The second principle provides a reasoning engine for deriving logical conclusions from logical assumptions. In this work we make a change in the first of these principles. Instead of the logical terms we apply unrestricted strings, which may contain variables and some nesting structure. We define and control senses by a unification of such strings, so that *string constituents, placed in similar positions of the unified strings, have similar senses*.

So, we compose logic clauses from the unrestricted strings, which may contain variable symbols and a nesting structure. We apply the resolution principle to such clauses in the same way it is applied in reasoning systems based on first order predicate calculi. Keeping an analogy with First Order Predicate Calculus (FOPC) terminology, we call the corresponding calculus First Order String Calculus (FOSC). FOSC is equivalent to FOPC when the class of strings, used in the clauses, is reduced to well-formed string representations of logical atoms.

The main advantage of this approach to logic is a complete elimination of logical terms and predicates from logical reasoning. Indeed, the need to design the artificial logical terms and predicates has obvious disadvantages. It is a complex intellectual work. The designed predicates may make the system biased and disregard important data features. Using FOSC, we can apply the classical logical reasoning methods without any artificial logical predicates at all. Instead of the predicates we use the patterns of unrestricted strings, which can be taken immediately from natural domain texts.

The patterns are produced using *alignment*, *generalization* and *structurization* of the original text fragments. They define the necessary formalization according to the regularities, applied via string *unification*. In this way we can produce an unbiased analysis of a text of unrestricted nature.

This work is related to cognitive informatics as follows. Jean Piaget [20] and his successors recognize that small children can generate and apply some kind of symbolic representation of the surrounding world and can expose a logically reasonable behavior. However, the children do not use explicit predicate notations. In order to make conclusions and conjectures, a child can apply *subconscious* versions of modus ponens and Church's substitution rule mentioned above. The child generally operates with *implicit* concepts, learned by examples, rather than with explicit concept definitions. This is a principal contradiction between the child's mind and the development of mathematics. In order to implement in our computer what a child can do easily and subconsciously, today we need a whole industry of manipulation with complex artificial notations! Consider formal and programming languages. Church's assumption is generally abided to, but the abundance of complex notations for explicit definitions often makes the relations between the algorithm components and their meanings hardly tractable.

The paradox we see here is that, when building computerized models, people tend to apply the *explicit* methods even where this leads to cumbersome and expensive models of simple phenomena. In order to understand this paradox, we try to understand what kind of formal elements *already exist* in unrestricted *natural* sources, such as natural language phrases and, in a future perspective, unknown codes, biologic sequences, images and other patterns of natural phenomena.

Exactly this objective leads us to the attempts to extract the regularities, contained in the raw domain texts, via the alignment and generalization of similar text fragments. FOSC can be considered *a knowledge representation tool, where all the knowledge is encoded using the patterns of raw text fragments treated as demonstrating examples*. The knowledge modeling and manipulation remains logic-based. The formal inference is based on the knowledge, extracted from similar text examples by alignment and generalization. In this way the *data* is transformed into *knowledge*. This can be considered a modeling of the internal knowledge in the brain. Indeed, in our brain we relentlessly align, match, compare, generalize and compose the images of external objects, contained in our memory. We still do not know how we do this. In FOSC we are trying to simulate these processes using the unrestricted text patterns, as described in Sections 2 and 3.

Generalization universe, introduced in Section 2, provides a special mathematical mechanism for extracting the knowledge, contained in a set of unrestricted strings, independently of any other knowledge. Although the introduced pair-wise string generalization and unification operations are not deterministic, we can consider a generalization universe as algebra with such operations. It can be considered a kind of concept algebra, where the concepts are *demonstrated* in string examples, but not formulated explicitly. Although we still do not specify the pair-wise string generalization algorithm (this can be done in many ways), the concept of generalization universe allows one to understand the expressive potential of the knowledge contained in the unrestricted texts. Indeed, we show (Section 3) that *any* algorithm can be encoded by alignment of suitable string examples. In this way we can study the inherent algorithmic content of a text independently of any other knowledge, and assess its cognitive complexity.

This is essentially an autonomic knowledge processing. The independence from other knowledge sources enables us to apply the inherent meaning of the unrestricted texts in a totally autonomic fashion. Furthermore, we can compose such meanings for the understanding of more complex texts, keeping the independence from any knowledge created by other means. This is discussed in Sections 4 and 5 (see also paper [6]), where we describe our experiments with FOSC, and in the concluding sections of the paper.

Some details of our experimental implementation of FOSC are described in Section 6 and in the Appendix.

2   Basic Definitions of First Order String Calculus

Let T be an infinite set of *terminal symbols,* V be an infinite set of *variable symbols, or variables,* $T \cap V = \emptyset$, and special symbols ¬ (not-sign), [and] (square brackets) do not belong to the sets T and V.

Finite strings, consisting of terminal symbols, are called *unstructured ground strings.* Unstructured ground strings and the results of operations of string structurization and string generalization, described below, are called *strings.* A *segment* of string S is a fragment of string S, which starts at some index i and ends at some index j, $i \leq j$.

Definition 2.1. A process and a result of inserting special square brackets [and] (in this order) into a string S are called *a structurization* of string S. That is, if S has a form $\alpha s \beta$, where α, s, β are segments, then $\alpha[s]\beta$ is the result of structurization. Here we assume that if segment s already contains a square bracket, then this segment contains the second bracket of the bracket pair, i.e. s is a valid string concerning its square bracket structure. Segments α and/or β may be empty.

Definition 2.2. A process and a result of substituting non-intersecting segments in a string S with variables are called *a generalization* of string S. Like in the previous definition, we assume that the substituted segments are valid strings. Also we assume that multiple occurrences of a single variable, introduced in a generalization of string S, are corresponding to equal segments of string S. The latest assumption is important for what follows; it is analogous to the assumption of uniqueness of values of mathematical functions. See examples at the end of this section.

Notation $f:S_1 \to S_2$ will stand for a generalization of string S_1 into string S_2. A sequence of segments of string S_1 (counted from left to right), which are substituted in the generalization, is called *a kernel* of the string generalization. A sequence of segments of string S_2, separated by the variables introduced in the generalization, is called *an image* of the string generalization. A generalization $f:S_1 \to S_2$ is called *reversible* if the corresponding backward substitution $f^{-1}:S_2 \to S_1$ is also a generalization; otherwise it is called *irreversible,* or *degenerate.* The result of sequential generalizations $f_1:S_1 \to S_2$ and $f_2:S_2 \to S_3$ is designated as $f_1f_2:S_1 \to S_3$ or $f_1f_2:S_1 \to S_2 \to S_3$.

Definition 2.3. A pair of string generalizations $\{f_1:S_1 \to S, f_2:S_2 \to S\}$ with the same image and the same variables, used for the substitution, defines *an isomorphism* between strings S_1 and S_2. (Here we assume that strings S_1 and S_2 do not contain common variables). String S is called *a common generalization* of strings S_1 and S_2. Segments of S_1 and S_2, related to the same variable in S, are called *synonyms* with respect to the string isomorphism. Synonyms take similar places in so aligned strings (see the examples at the end of this section). So defined string isomorphism is called *most specific* if it can not be represented in the form $\{f_1h:S_1 \to S' \to S, f_2h:S_2 \to S' \to S\}$, where $h:S' \to S$ is a degenerate generalization. In this case string S is called *a most specific common generalization* of strings S_1 and S_2. See examples in the final parts of this section and in Section 4.

Operation of string generalization, considered a relation $g(s_1, s_2)$: "string s_2 is a generalization of string s_1", forms a preorder on the set of all strings defined above. The maximal elements in this set are strings, formed by single variables. The minimal elements are ground strings.

Definition 2.4. The result of simultaneous substitution of all occurrences of a variable in a string S with some string is called *an instantiation* of string S. We assume that such substitution can be done simultaneously for one or more variables of the string S.

String generalization and instantiation are inverse relations of each other. Strings S and S′ are called *equivalent* to each other if they are instantiations of each other. String equivalence should not be confused with string isomorphism, in which the strings are aligned in order to produce a common generalization.

String equivalence forms a relation of equivalence on the set of all defined strings. Equivalent strings can be transformed to each other by a reversible generalization. Speaking of strings with variables we usually bear in mind the corresponding equivalence classes. Relation of preorder $g(s_1, s_2)$ between strings becomes a relation of partial order between the classes of equivalent strings.

It can be easily shown that any string generalization can be represented as a most specific common generalization of a pair of ground strings, which have some similarity. The non-matching fragments of so aligned strings become synonyms with respect to the corresponding string isomorphism. (See examples in the final part of Section 4).

Definition 2.5. Given a set E of ground strings, its closure with respect to the operation of taking most specific common generalization, factorized by the relation of string equivalence, is called *a generalization universe* G(E) of set E. (See example in Fig. 1 and in the table that follows). The elements of generalization universe G(E) correspond to all possible regularities, which can be discovered by multiple pair-wise alignments of elements of set E. More specific regularities correspond to the lower elements in the ordering of string generalization. This concept of regularities is discussed in more detail below.

Definition 2.6. Given two strings S_1 and S_2, consider a possibility to create a common instantiation S of these strings. String S and the corresponding substitution of variables (if possible) are called *a unification* of strings S_1 and S_2.

Definition 2.7. For any string S, *a string literal* (or simply a literal), produced from S, is defined either as S or ¬S. (Remind that not-sign ¬ does not belong to the sets of symbols T and V). Keeping an analogy with mathematical logic terminology, we call S and ¬S *a positive* and *a negative* literal respectively.

Definition 2.8. A pair (H, B), where H is a string and B is a finite set of string literals, is called *a string clause,* or simply a clause. Set B can be empty. Keeping an analogy with Horn clauses, considered in Logic Programming, we call H and B *a clause head* and *a clause body* respectively. String clause (H, B) is called positive if its body B contains only positive literals.

A notation H :– B will be applied for the designation of a string clause with head H and non-empty body B. A clause with head H and empty body will be designated simply as H. A unification of clause head H of a string clause H :– B with string S of a literal S or ¬S is called *an instantiation of the string clause* by this literal. We assume that the substitution of the corresponding variables is done globally in the entire clause H :– B, including its body.

Definition 2.9. A finite set of string clauses is called *a string theory, or FOSC theory.* We say that a string theory P is *composed* of strings, applied in its literals.

Definition 2.10. Assume that an instantiation of a positive string clause (H, B) by a ground string α makes every string $S \in B$ unifiable with some string $H_{\alpha,S}$. Clause (H, B) is called *a description clause* for the ground string α. String α is called *the supporting example;* string H is called *bondable* by the set B; elements of set B are called the *bounds*; string H is also called *the acceptor* of the bounding; string $H_{\alpha,S}$ is called *the attractor* for bound S.

Our terminology of acceptors and attractors will be clarified in Section 7.

Examples. In our examples we apply the following notation. Variable names and terminal symbols in strings are separated by a white space and/or square brackets. Variable names start with capital Roman letters and may contain only Roman letters and digits. The terminal symbols are designated by sequences of characters and letters (excluding white space, the neck-symbol :–, double dots, double commas and characters ¬, [,]) and do not start with capital Roman letters. Like in Prolog language, proper names, considered as terminal symbols, are written non-capitalized. Strings in a clause body are separated by double commas and are terminated by a double dot at the end of the clause. Program comments start with the percent sign %.

Consider strings S_1, S_2, S, S', defined as follows:

S_1 = "[a , b] $\Rightarrow$ [b , a]", S = "[A , B] $\Rightarrow$ [C , A]",
S_2 = "[x , y , z] $\Rightarrow$ [z , y , x]", S' = "[A , B] $\Rightarrow$ [B , C]".

These strings contain distinct terminal symbols a, b, x, y, z, comma, $\Rightarrow$, and distinct variable symbols A, B, C. Remind that square brackets [and] are special symbols, which define a nesting structure on strings. Strings S_1 and S_2 demonstrate a reversing of comma-separated lists of entities. String S realizes a regularity, which may be verbally formulated as follows: the first element of the first comma-separated list in square brackets is equal to the last element of the second comma-separated list in square brackets. String S' realizes a slightly different regularity.

A generalization $f:S_1 \rightarrow S$ is defined by substituting a with A, 1st occurrence of b with B, 2nd occurrence of b with C. A generalization $g:S_2 \rightarrow S$ is defined by substituting x with A, "y , z" with B, "z , y" with C. Pair {f, g} defines an isomorphism between strings S_1 and S_2. String S is a common generalization of strings S_1 and S_2.

A generalization $f':S_1 \rightarrow S'$ is defined by substituting 1st occurrence of a with A, b with B, 2nd occurrence of a with C. A generalization $g':S_2 \rightarrow S'$ is defined by substituting "x , y" with A, z with B, "y , x" with C. Pair {f', g'} defines another isomorphism for strings S_1 and S_2. String S' is another common generalization of strings S_1 and S_2.

Strings S and S' are not equivalent, so we find that a pair of strings may be generalized into two essentially different, i.e. non-equivalent and not generalizing one another, common generalizations.

The following string alignments define two essentially different unifications of string S with string [c a t , d o g , m o n k e y] $\Rightarrow$ [W]:

[	A	,	B	] $\Rightarrow$ [	C , A	]
[	c a t	,	d o g , m o n k e y	] $\Rightarrow$ [	W	]

[	A	,	B	] $\Rightarrow$ [	C , A	]
[	c a t , d o g	,	m o n k e y	] $\Rightarrow$ [	W	]

The alignments differ in the matching of commas. The unifications are defined by the substitution of variables A, B and W:

[c a t , d o g , m o n k e y] $\Rightarrow$ [C , c a t]   (from the 1st alignment),
[c a t , d o g , m o n k e y] $\Rightarrow$ [C , c a t , d o g]   (from the 2nd alignment).

The following strings are instantiations of the first unification:

[c a t , d o g , m o n k e y] $\Rightarrow$ [m o n k e y , d o g , c a t],
[c a t , d o g , m o n k e y] $\Rightarrow$ [j a b b e r w o c k y , c a t] .

The following strings are instantiations of the second unification:

[c a t , d o g , m o n k e y] $\Rightarrow$ [m o n k e y , c a t , d o g],
[c a t , d o g , m o n k e y] $\Rightarrow$ [j a b b e r w o c k y , c a t , d o g].

In Fig. 1 a fragment of the generalization universe, created from a set of ground strings containing strings S_1 and S_2, is depicted. It contains strings S and S'. Note that the depicted strings are representatives of the corresponding equivalence classes. For instance, the class of equivalent strings, which is represented by string [A , B] $\Rightarrow$ [C , A], contains string [X , Y] $\Rightarrow$ [Z , X], where X, Y, Z are distinct variables. The less general elements of the generalization universe realize more specific regularities.

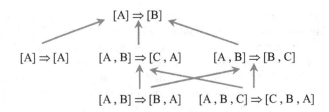

Fig. 1. A fragment of generalization universe of the set of ground strings "[a] $\Rightarrow$ [a]", "[b] $\Rightarrow$ [b]", "[a, b] $\Rightarrow$ [b, a]", "[x, y] $\Rightarrow$ [y, x]", "[a, b, c] $\Rightarrow$ [c, b, a]", "[x, y, z] $\Rightarrow$ [z, y, x]". The arrows show the relation of string generalization.

Below is a simple string theory example:

Example 2.1. A string theory of reversing comma-separated lists

[A , B] $\Rightarrow$ [C , A] :– [B] $\Rightarrow$ [C]..
[A] $\Rightarrow$ [A]..

This is a theory of reversing comma-separated lists of lists of terminal symbols. The application of a string theory to a string is defined in the following section (see Algorithm 3.1). Remind that $\Rightarrow$ is an ordinary terminal symbol, and that string clauses in a string theory are terminated by a double-dot.

The application of the above theory to the string [c a t , d o g , m o n k e y] $\Rightarrow$ [W] will produce the string [c a t , d o g , m o n k e y] $\Rightarrow$ [m o n k e y , d o g , c a t]. This is done recursively as follows. The input string (goal) is unified with the clause head [A, B] $\Rightarrow$ [C, A], forming a new sub-goal [d o g , m o n k e y] $\Rightarrow$ [C] from the clause body [B] $\Rightarrow$ [C]. In the next recursion step, a sub-goal [m o n k e y] $\Rightarrow$ [C] is similarly formed. This sub-goal can be unified only with the clause [A] $\Rightarrow$ [A], resulting in string [m o n k e y] $\Rightarrow$ [m o n k e y]. The output string is then formed by the variable substitutions on all recursion levels. Note that in this theory the 2nd possibility of aligning text [A, B] $\Rightarrow$ [C, A] with the goal is not applied.

This example demonstrates an incorporation of parsing facilities immediately into a logical reasoning, *avoiding the concepts of predicate and formal grammar.* Some very important implications of this avoiding are discussed in Section 4 in relation to natural language syntax and semantics.

Consider Definition 2.10. The first clause of Example 2.1 is a description clause of ground example [a , b] $\Rightarrow$ [b , a]. String [A , B] $\Rightarrow$ [C , A] is the acceptor of the corresponding bounding; string [A] $\Rightarrow$ [A] is the attractor of the bound [B] $\Rightarrow$ [C].

Now, the same clause is a description clause of ground example [a, b, c] $\Rightarrow$ [c, b, a]. String [A, B] $\Rightarrow$ [C, A] is simultaneously the acceptor and attractor of this bounding.

In the following example, a fragment of generalization universe G(E) of a set of English sentences is shown. Sentence numbers in the columns mark the sources for building most specific string generalizations. The most general of these generalizations is underlined near the bottom of the table.

Sentence No.	Set of English sentences	Sentence No.	Generalizations derived from 2 sentences
1	the crayon is found by mary	1,2	the X is found by mary
2	the pencil is found by mary	1,3	the X is Y by mary
3	the book is stolen by mary	1,5	the X is found by Y
4	the book is returned by mary	3,4	the book is X by mary
5	the book is found by bill	3,5	the book is X by Y
6	the book is found by john	5,6	the book is found by X
Sentence No.	Generalizations derived from 3 or more sentences		
(1,2),(1,3)	the X is Y by mary		
(1,2),(1,5)	the X is found by Y		
(1,2),(3,5)	the X is Y by Z		
(3,4),(5,6)	the book is X by Y		

3 Algorithmic Form of a FOSC Theory

We introduced FOSC, abandoning the concept of predicate and trying to keep other characteristics of FOPC, related to logic programming. Now we are going to apply FOSC as a means for creating algorithms. *From now on we consider the string theories and the bodies of string clauses as sequences.* Any Prolog program can be considered a string theory.

A string theory defines an algorithm for string processing as described below. The algorithm is defined recursively, keeping an analogy with the application of a Prolog program P to a Prolog goal S. String unification algorithm, applied here, is a simple modification of the common logic term unification algorithm (see Robinson, [22];

Lloyd, [15]), which is specified for working with the unrestricted strings with a nesting structure and variables. The details of this algorithm are not important for the understanding of what follows; interested readers can find them in the Appendix.

Algorithm 3.1. Apply a string theory P to a string-goal S.

1) Starting from the top of P, find the first clause H :– B, not considered before, such that H can be unified with S. Return an unsuccessful status and exit if this is impossible.

2) Build a clause H_1 :– B_1 as the first possible instantiation of clause H :– B by string S, not considered before. Proceed to Step 1 if this is impossible (that is, all such instantiations are already considered).

3) If B_1 is empty then return H_1 as the successful result of the algorithm and exit. Otherwise sequentially apply P to the strings of all positive and negative literals of B_1. In case of success, return H_1 as the successful result of the algorithm and exit. In case of non-success proceed to Step 2.

In Step 3 we assume that string H_1 is modified by global substitutions of its variables (if any), made during the corresponding unifications. Also we assume that the application of P to any negative literal returns an unsuccessful status. The latest assumption is related to the closed-world assumption (CWA), which states that what cannot be deduced from a theory is considered wrong. The application of P to a negative literal is attempted only if the string of this literal is a ground string (this is so called safeness condition). This algorithm is based on SLDNF-resolution rule; see [15].

It is now the best time to consider Example 4.1, where we analyze the application of a natural language related string theory to natural language related goals.

The representation of algorithms as string theories is Turing-complete, which means that we can represent any algorithm as a string theory. In order to prove this we will prove that any normal Markov algorithm [17] can be represented as a string theory. This is stated in the following theorem.

Theorem 3.1. Let M be any Markov algorithm, written in a base alphabet B and a local alphabet L, where $(B \cup L) \subset T$ and T is the set of terminal symbols applied in string theories. There exists a string theory P_M, which, given a goal $[\xi t\xi] \Rightarrow [W]$ (where t is a string, ξ and $\Rightarrow$ are terminal symbols not belonging to alphabets B and L, W is a variable), does the following:

- *Transforms the goal into a string $[\xi t\xi] \Rightarrow [\xi t'\xi]$ and stops if M transforms string t into a string t' and stops;*
- *Produces a string $[\xi t'\xi] \Rightarrow [[fail]]$ and stops if no rule of M is applicable to some derivation t' of t, produced by M;*
- *Never stops if string t leads M to an infinite application of the Markov rules.*

The theorem can be proved simply by rewriting the rules of Markov algorithm $\alpha \rightarrow \beta$ (normal rules) and $\alpha \rightarrow .\beta$ (stop rules) in the form of FOSC clauses $[X\alpha Y] \Rightarrow [W]$:– $[X\beta Y] \Rightarrow [W]$ and $[X\alpha Y] \Rightarrow [X\beta Y]$ (a clause with empty body) respectively. Here X, Y and W are FOSC variables. Clause $[X] \Rightarrow [[fail]]$ (also a clause with empty body) is added at the end of the theory in order to prevent backtracking.

Representation of Markov algorithms, written in other alphabets, can be done by encoding, using the set T of terminal symbols. So, for any Markov algorithm we can construct a FOSC theory with a similar behavior. Now, any Turing machine can be represented as a Markov algorithm [17], so we can conclude that any Turing machine can be represented as a string theory and that the possibility of representing algorithms in the form of string theories is also Turing-complete. The same result can be proven in a different way, by observing that Algorithm 3.1 subsumes context-sensitive string rewriting.

Theorem 3.1, along with the means for pair-wise string generalization and unification, described above, leads to an interesting treatment of Church-Turing thesis. For a detailed description of this thesis, see e.g., [12]. In a common formulation the thesis states that every effective computation can be carried out by a Turing machine. The thesis contains the following requirements about an algorithm (or a method):

1). The method consists of a finite set of simple and precise instructions, which are described with a finite number of symbols.
2). The method will always produce the result in a finite number of steps.
3). A human being with only paper and pencil can in principle carry out the method.
4). The execution of the method requires no intelligence of the human being except that which is needed to understand and execute the instructions.

Instructions, mentioned in p.1, usually are organized in a form of formal algorithm description. Some representation of the input data for the algorithm is also assumed.

String theories provide an alternative view. We do not need the *instructions*, mentioned in p.1. Instead, we use a finite set of *data examples* (in a string form), produce the generalized patterns of these examples, and combine the patterns using a universal set of rules, which are described in Algorithm 3.1.

Data example patterns now become a fundamental part of the algorithm. So we came to the following interpretation of Church-Turing thesis:

Thesis 3.1. *Every effective computation can be carried out using sequences of generalized patterns of data examples and a universal set of rules for combining such sequences.*

Indeed, any string theory consists of clauses, each of which is a sequence of generalized string examples. The universal set of rules for combining such sequences is defined in Algorithm 3.1.

The set of instructions now consists of the following two distinct subsets: a) a fixed set of universal rules, determining how to apply a string theory to a string; b) a finite set of data generalizations, which now play the role of instructions, or model examples, for the processing of another data. We especially emphasize that the set of *actions* (instructions of type (a)) here is fixed forever. All other means for creating algorithms are essentially *data*, which can be extracted from the unrestricted text fragments by a generalization.

Now, any such data generalization can be derived from a pair of demonstrating examples, which have some similarity, using a most specific common generalization. So we come to the following additional interpretation of Church-Turing thesis, where the process of generalization itself is a part of the computation:

Thesis 3.2. Every effective computation can be carried out using sequences of data example pairs and a universal set of rules for combining such sequences.

In more general terms we can reformulate both theses as follows:

Thesis 3.3. Every possible computation can be carried out by alignment and matching of some data examples.

As an important implication of this thesis, we can build any algorithm without specially designed instructions for the definition of actions, logic conditions, relations, data structures, properties, procedures etc. Everything is *automatically* defined by the regularities, contained in the generalized string examples. Likewise, we can build a formal theory of any data, presented in a string form, without any previously defined mathematical concepts, grammars, or predicates related to the data. Every concept is *implicitly* defined by the corresponding variable, which represents the non-matching string constituents aligned for producing the string generalization. In Sections 4 and 5 we provide examples and discuss this implicit definition of concepts in more detail.

When we *think* of an algorithm, we usually think of steps, which are applicable to similar data examples. Now we see that we can *define* algorithms and theories, using only data examples, which have some similarity. Any known algorithm or formal theory has a prehistory of finding similarities in data, which may become a part of this algorithm or theory if we reformulate it according to Thesis 3.3.

This observation is consistent with our intentions to simulate the internal knowledge processing in the brain. According to our hypothesis, described in the introduction section, the alignment, matching, generalization and composition of string examples into clauses in FOSC simulate the way our brain manipulates with the images contained in our memory.

4   Case Study: Implicit Syntax and Semantics of Natural Language Text as a FOSC Theory

Recall once more the assumption by Alonzo Church, mentioned in our introduction. In Church's context the mathematicians manually compose the *complex names*. Every applied predicate and functional symbol should be explicitly declared and defined. The senses are defined *explicitly*.

Instead of doing this, we can align and generalize sample strings, where the senses are *demonstrated*, and then apply these senses using the generalized string patterns. Text constituents, related to each other in the string alignment, define the senses *implicitly*. It is important to realize that we extract and operate with these senses without the explicit formal definitions. We just *align* some sample strings, *assume* that this alignment is correct, make the generalization, and then use so introduced variables in clauses via clause instantiation. Like explicit formal definitions, such an implicit definition of senses can be correct (if our assumptions about the string alignments are sound) or incorrect (if they lead to a wrong inference). Clauses, composed of the generalized strings, serve the role of axioms for deriving the conclusions.

Let us apply this framework for the definition and manipulation of senses of natural language phrases. A FOSC theory can be considered as ontology in the following way:

- Word occurrences in clause heads are considered the concept definitions;
- Word occurrences in clause bodies are considered the suggested concept applications;
- Semantics of the words and word combinations is defined by the suggested concept applications.

Consider the problem of question answering. Assume that there exists a set of simple English sentences, not limited by any specific grammar, which describe certain facts. (See the facts in Example 4.1; the structurization of these sentences will be discussed later). The facts describe a simple situation, which may happen in a real-world library or store.

Note that we *intentionally* do not apply any linguistic concepts: everything will be defined implicitly. We would like to develop a question answering system, which answers questions about the facts. Also we would like our system to be easily expandable by new facts, presented just in a simple English, and new kinds of questions.

Some common approaches for formalizing the involved NLP phenomena are based on abstruse explicit formalisms, which include formal grammars, first order logic with numerous predicates, graph-based representations of semantics, large thesauri with semantic tags etc. Using FOSC we can simplify this. In Fig 2 we depict the difference between the application of FOSC to this NLP problem (see the white rectangles only) and the approach based on FOPC combined with a formal grammar formalism (see *all* the rectangles, assuming a predicate-logic form of the inference engine, ontology and the formal model).

We now discuss two implementations of the same algorithm for question answering, which are created according to the paths described in Fig. 2: path 1-2 and path 3-4-5-6-7-8. The FOSC implementation (path 1-2) is given in full details. The FOPC implementation, which is much more complex, is described briefly.

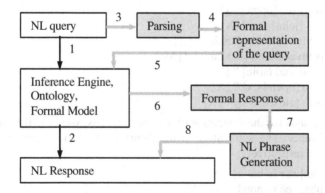

Fig. 2. Comparison of FOPC and FOSC approaches to question answering: ⟶ *FOPC approach* assumes the application of formal grammars and leads to the need to design predicates, grammars, a semantic model, and implement a parser, a formal reasoning engine, and a NL phrase generator. ⟶ *FOSC approach:* Everything is defined via the immediate interaction of generalized patterns of NL phrase examples.

FOSC theory, presented in Example 4.1, forms the answers according to the meaning of natural language phrase patterns, contained in its clauses. In this example we intentionally combine a linguistic type of inference with a logic inference, related to the subject. The theory contains a set of inference rules, followed by some facts in a simple English form. Note the negation signs ¬ in the 5th and 8th clauses. Note also that the scope of any variable is limited by the clause containing this variable.

Example 4.1. A FOSC theory for question answering.

> can [A] see [X] ? :– [A] can see [X]..   *% Inference rules*
> what can [A] see ? [X] :– [A] can see [X]..
> what is in [A] ? [X] :– [X] is in [A]..
> what is visible on [A] ? [X] :– [X] is visible on [A]..
> what is invisible on [A] ? [X] :– [X] is on [A] ,, ¬[X] is visible on [A]..
> what is on [A] ? [X] :– [X] is on [A]..
>
> [A] can see [X] :– [A] is standing near [B] ,, [X] is visible on [B]..
> [A] is visible on [B] :– [A] is on [B] ,, ¬[A] is in [closed X]..
> [A] is standing near [B] :– [A] has approached [B]..
> [X] is on [A] :– [X] is in [B] ,, [B] is on [A]..
>
> [a book] is in [open box]..   *% Facts*
> [a notebook] is in [closed box]..
> [open box] is on [red table]..
> [closed box] is on [red table]..
> [john] has approached [red table]..

Consider the following input queries for the theory:

> 1) can [john] see [a book] ?
> 2) can [john] see [a notebook] ?
> 3) can [john] see [open box] ?
> 4) what can [john] see ? [X]
> 5) what is visible on [red table] ? [X]
> 6) what is invisible on [red table] ? [X]
> 7) what is on [red table] ? [X]
> 8) what is in [open box] ? [X]
> 9) what is in [closed box] ? [X]

Variable X stands for the fragments of the query results, which will be instantiated during the inference. The application of the theory to the above inputs provides the following results respectively:

> 1) can [john] see [a book] ? *Yes.*
> 2) can [john] see [a notebook] ? *No.*
> 3) can [john] see [open box] ? *Yes.*
> 4) what can [john] see ? [open box]. *Yes;*
>   what can [john] see ? [closed box]. *Yes;*
>   what can [john] see ? [a book]. *Yes.*
> 5) what is visible on [red table] ? [open box]. *Yes;*
>   what is visible on [red table] ? [closed box]. *Yes;*
>   what is visible on [red table] ? [a book]. *Yes.*

6) what is invisible on [red table] ? [a notebook]. *Yes.*
7) what is on [red table] ? [open box]. *Yes*;
   what is on [red table] ? [closed box]. *Yes*;
   what is on [red table] ? [a book]. *Yes*;
   what is on [red table] ? [a notebook]. *Yes.*
8) what is in [open box] ? [a book]. *Yes.*
9) what is in [closed box] ? [a notebook]. *Yes.*

Words *Yes* and *No* in Italic here indicate a successful or unsuccessful status of the application of the theory to the query. Some queries have multiple answers. The negation in query (2) is done according to the closed-world assumption rule. Remind that square brackets define nesting structures on the applied string patterns and constrain the unification.

Essentially, this theory contains some algorithmic knowledge about syntax and semantics of English words and phrases, and allows us to apply this knowledge. Curiously, we did not try to define this algorithmic knowledge explicitly, using previously formalized artificial lexical, syntactical or semantic concepts. *We apply this algorithmic and linguistic knowledge immediately by examples!*

The theory expresses and supports reasoning about common relations between objects (a book, a notebook, john etc). Among the relations we find *inclusion relations* "is in", *topological relations* "is on", "is standing near", *time-related relations* and *action patterns* "[A] has approached [B]" and relations concerning a visibility of objects and an ability of a person to perceive objects. What is important for us, the relations are defined *implicitly*, via the generalized examples of their usage. The roles and types of the objects-operands of the relations are also defined implicitly, according to the word positions in the phrase patterns.

We could define the relations *explicitly* through a conventional artificial FOPC form (e.g., a Prolog form) by inventing the following predicate notations:

    is_in (a_book, open_box).
    is_in (a_notebook, closed_box).
    is_on (open_box, red_table).
    is_on (closed_box, red_table).
    has_approached (john, red_table).

If we try this, we will find that we need to invent additional explicit notations for the relations and to create the corresponding axioms, e.g.

    can_see (A, X) :– is_standing_near (A, B), is_visible_on (X, B).
    is_visible_on (A, B) :– is_on (A, B).
    is_standing_near (A, B) :– has_approached (A, B).

Such predicates and axioms would define the FOPC semantics according to arrows 5 and 6 in Fig. 2. However, we will also have to build a grammar for parsing the input queries 1-9 and for transforming them into the predicate form (see arrows 3 and 4). Then we will have to implement the corresponding parser. Then we will need another grammar and a special phrase generator for generating a natural language representation of the inference results (see arrows 7, 8).

Trying to develop this FOPC-based formalization, we will find that the simplicity, readability, independence from other formalisms, and the flexibility for developing the theory are lost forever. This is a confirmation of our complexity observation, related to the assumption by Alonzo Church, given in the introduction section.

Using FOSC we can avoid these complications (see arrows 1, 2 in Fig. 2): in our theory the artificial concepts of formal grammar and logical predicate are not applied at all. So, FOSC provides an alternative mechanism for the formalization of text structure and meaning, as well as for text analysis and generation.

Although we avoid the predicate notations, our theory is still *a formal theory*, conforms to Church's assumption, and supports a logical reasoning, which can be applied to other objects, phrase patterns and lexemes. We not only exploit the *syntax* of the involved phrases. We exploit the inherent *semantics* of the English words and word combinations "can see", "is in", "is visible", "has approached" etc, without having to create the explicit formalization of this semantics. In this way we can take advantage of the syntax and semantics of any natural text fragment without the need to define them explicitly.

String patterns, applied in Example 4.1, can be created as follows. The following table contains a verbal description of the meaning of several string pairs, composed of similar English sentences and dialogue samples:

String example pair	Verbal description
can john see a book ? can mary see open box ?	Question about the visibility of some object or about the ability of a person to perceive objects
john can see red table mary can see closed box	Statement that some object is visible or that some person can perceive objects
what can john see ? a book what can mary see ? open box	Question and answer about the visibility of some object and about the ability of a person to see this object
what is in open box ? a book what is in room ? red table	Question and answer about a containment

When we have such string pairs, we do not need to *write* a theory. We can *build* it, manually or automatically, from the string pairs, using structurization and most specific common generalization (cf. Section 2) as follows:

String example pair	Structured most specific common generalization
can john see a book ? can mary see open box ?	can [A] see [X] ?
john can see red table mary can see closed box	[A] can see [X]
what can john see ? a book what can mary see ? open box	what can [A] see? [X]
what is in open box ? a book what is in room ? red table	what is in [A] ? [X]

The *implicit definition* of the semantics assumes such an abduction of the semantics from the natural language text samples, where this semantics is *demonstrated*, rather than an explicit mathematical definition what this semantics is. In terms of the assumption by Alonzo Church, we create the *complex names* for a logic formalization of a problem immediately from the generalized patterns of natural texts, belonging to the problem domain, rather than from the artificially constructed logical atoms and predicates. We create an algorithm *by incorporating the demo examples immediately into the algorithm description*, according to Thesis 3.1.

Example 4.1 is a "toy" reasoning system, which only demonstrates that we can do without the *explicit* frameworks, such as formal grammars and predicates. In this work we only show that FOSC can be applied as an alternative to these techniques, enabling us to avoid some of their problems. We did not discuss the automatic acquisition of inference rules for specific NLP tasks: we relate this to a future research and experiments. Note that some modern corpora and machine learning methods for building large grammars and transfer bases can be similarly applied for building FOSC rules.

Some problems of ambiguity and variability in natural languages can be tackled by the acquisition and alignment of phrases, which have similar structure and meaning, as shown in our examples. As an interesting example of avoiding ambiguity, consider the word "can", contained in our examples. Example 4.1 implements reasoning, related to a treatment of this word as a modal verb. Due to the constraints, imposed by unification of the phrase patterns, containing this word, this reasoning cannot be applied when this word is correctly treated as a noun. So, *the ambiguity is simply avoided, just like we avoid it in our speec*h. We did not exert any efforts for resolving this ambiguity besides the application of correct phrase patterns: the ambiguity simply does not arise.

Now let us recall the questions, asked in the introduction section. Our examples show the following.

Conclusions from the case study. We can abide to Church's assumption without *artificial* symbolic notations: we apply the notations, which already exist in *natural* sources. We only need to make the generalization and structurization of the natural strings, such as natural language phrases, suitable for our purposes.

We can apply natural phrases and their meanings immediately in algorithm definition. In order to make inferences from a natural language text, we do not need a *prior* semantic knowledge, expressed mathematically. We only need to supply our algorithm with the patterns of correct natural language reasoning.

We can use natural language words and concepts in a computer program almost as fluently as in human speech. We only need to make patterns of the phrases, where these words and concepts are correctly demonstrated, and include them into our program.

We can apply patterns of unrestricted texts immediately in reasoning, consistently with the meaning of these patterns. According to the method of implicit semantics, we can define and manipulate with this meaning implicitly, without reference to any mathematical construction known a priori.

So, we need to supply our algorithms with the generalized and structured examples of correct reasoning. We relate to a future research the methods of automatic extraction and composition of such example patterns into the theories for performing specific tasks. This is discussed in Sections 6 and 7.

The principal limits of an *artificial* formalization, such as that depicted in the path 3-4-5-6-7-8 of Figure 2, lie in the need to invent and accommodate the heterogeneous and complex artificial formalisms. Indeed, the more diverse and complex constructions we invent, the harder are these constructions to combine. Thesis 3.1 and the method of implicit definition of senses provide some hope that these limits can be overcome by replacing the artificial mechanisms with the clauses composed from the generalized examples of *natural* texts, where the senses are already combined in a natural way.

An interesting implication of this is the possibility to implicitly define and apply the meaning of idiomatic expressions. We consider this a perspective research direction for natural language processing. Other interesting directions include: 1) Generating a large question answering system from question and answer examples and a background knowledge expressed in a text form; 2) Creating an encyclopedia of *working* algorithms, which are described in common-sense verbal terms, as we do in the following section.

5 Case Study: FOPC Specifics as a FOSC Theory

The following string theory example (Example 5.1) demonstrates an algorithm for logic formula transformation where some new object variables of a *predicate* calculus are automatically generated. Consider the following well-known rule for transformation of formulas in a first-order predicate calculus with quantifiers:

Formula (Qx)F $\vee$ (Qx)H of the calculus can be transformed into formula (Qx)(Qy) (F $\vee$ G), where:

Q is a quantifier $\forall$ or $\exists$;
Formulas F and H do not contain variable y;
Formula G is the result of global replacing of variable x in formula H with variable y.

Example: formula $(\forall x)a(x,y) \vee (\forall x)b(x,t)$ can be transformed into formula $(\forall x)(\forall z)$ $(a(x,y) \vee b(z,t))$.

This is essentially an *informal* description, which contains some *formal* elements in a textual form. We do not intend to study this predicate calculus: we use the above description only for demonstration of our methods of extracting algorithms *immediately* from string examples, which may contain formal and informal fragments.

This description defines a transformation of logic formulas. The following string theory does this transformation using only the generalized string examples of this transformation as a building material. A working Sampletalk version of this theory can be found in our website (see below) along with an experimental implementation of Sampletalk compiler. Note the prefix "¬" in the 4th clause, which denotes a negation as failure.

Example 5.1. A string theory for logic formula transformation.

% Goal. Variable W stands for the result of the transformation:
the result of shifting quantifiers in formula $(\forall x\ 0) [a (x\ 0, y)] \vee (\forall x\ 0) [b (x\ 0, t)]$ is formula W ..

% The main theory clause:
the result of shifting quantifiers in formula (Q X) [F] ∨ (Q X) [H] is formula (Q X)
(Q Z) ([F] ∨ [G]) :–

X is notation for object variables ,,
Z is notation for object variables ,,
formula F does not contain object variable Z ,,
formula [G] is the result of replacing X by Z in formula [H] ..

% Definition of the variables for the predicate calculus:
x 0 is notation for object variables ..
X 1 0 is notation for object variables :–
X 0 is notation for object variables ..

% A straightforward explanation what is "does not contain" and "contains":
formula F does not contain object variable Z :–
¬word F contains word Z ..
word A X B contains word X ..   *% Note free FOSC variables A and B here*

% Algorithm for replacing variables in the predicate-logic formulas:
formula [A Y N] is the result of replacing X by Y in formula [A X M] :–
formula [N] is the result of replacing X by Y in formula [M] ..
formula [A] is the result of replacing X by Y in formula [A] :–
formula A does not contain object variable X ..

This theory produces the following output: the result of shifting quantifiers in formula (∀ x 0) [a (x 0, y)] ∨ (∀ x 0) [b (x 0, t)] is formula (∀ x 0) (∀ x 1 0) ([a (x 0 , y)] ∨ [b (x 1 0, t)]). Remind that square brackets [and] are special symbols (not terminal symbols), which define a nesting structure on strings and constrain the possibilities of string unification. Sometimes there are too many possibilities of unifying texts with variables. The nesting structure constrains this.

Constructing such a theory, we do not have to think of the arrays, formal grammars, parsing, loops, logical conditions and procedures, necessary in a traditional programming. We do not have to think of logical relations, too, although we build string clauses with a logical behavior. We only think of and deal with string examples. The above theory can be similarly developed into a more sophisticated logic formula transformation algorithm.

In a way, this example shows a hybrid approach, where we *implicitly* define and apply a strict formal calculus syntax via the generalized examples that only *demonstrate* this calculus. In a similar and seamless way we can embed any system of formal notations into the example-based reasoning, according to Thesis 3.1. So we can combine the notations of any formal theory or ontology, which is somewhere defined *explicitly*, with natural language reasoning.

6   Experiments with Sampletalk Compiler

Sampletalk language [6] is an implementation of FOSC. An experimental Sampletalk compiler, along with the working examples described here, currently can be downloaded from website www.sampletalk.com. In this compiler we apply a convention that the string segments, used for the substitution of variables in text unification, cannot be empty. The compiler is implemented using SWI-Prolog and includes some Prolog specifics: machine-oriented constants, a cut operator etc. Currently only this experimental implementation of FOSC exists. About 30 working examples of Sampletalk programs can be found in our works (see [6] and the references therein). A large Sampletalk program for an experimental linguistic application can be found in [24].

Sampletalk (FOSC) Programming vs. Prolog (FOPC) Programming. Our examples show several essential differences between Sampletalk programming, based on the alignment of generalized strings, and Prolog programming, based on the artificial predicate notations.

First of all, we do not need the embedded predicates *append/3, member/2, arg/2, var/1, atom/1, univ/2* etc, applied in Prolog for the analysis and transformation of the data between different representations (lists, predicates, clauses, strings etc). These predicates, just like the basic operators of any conventional programming language, form the basis of the *artificial* formalization in Prolog programming. In Sampletalk the roles of the elementary operators play these more expressive and intuitive *natural* expressions: *is, in, by, the result of, does not, for* etc. What is important for us, the interaction of program clauses in Sampletalk can be governed by such expressions consistently with their natural common-sense meaning. Furthermore, we can apply as many natural expressions as we want. This is impossible in Prolog.

The second difference is that, given the informal problem description and the examples of desired processing, *we do not invent or write the program code*. We *embed* the examples and the description fragments immediately into the program, trying to preserve their text form as much as possible. (Actually, we do only string generalization, structurization, and some adaptation of the texts in order to comply with FOSC notation). This is also impossible in Prolog, unless the problem in question is already described in a predicate form.

As the other side of the coin, in Sampletalk we need a strict matching of the string patterns, which a priori are unrestricted. This may impose a problem if the patterns contain commas, capitalizations and other variations, which often are unnoticed when a human deals with a natural language text. Prolog compilers do not allow such syntax freedom and prevent the programmer from inserting the inconsistent symbols into a program code. In most cases this problem can be easily overcome by a preprocessing of the text in question, as we did in our examples. Some other differences between Sampletalk and Prolog programming are discussed in Section 4 and in paper [6].

7 Discussion

This work contains only an initial investigation of the possibility to analyze unrestricted texts via matching of the text fragments. Some questions, raised by this work, can be answered only regarding specific applications, e.g., natural language processing systems. Here we do not describe any complete system of this kind and do not compare our approach to others in terms of performance, coverage and disambiguation. Among the questions, we should ask the following: How to choose text fragments to be considered as the strings? How to automate the generalization and structurization? How to choose the string pairs to build a theory from? How to generate a FOCS theory with a desirable behavior? Can we apply the existing example-based machine learning methods in order to build a FOSC theory?

Here we only outline some useful directions. More substantial answers to these questions are related to a future research.

Extracting a String Theory from String Examples. FOSC theories can be extracted from string examples using some machine learning methods. We already know that the generalized string patterns can serve as a building material for constructing such theories. Using simple statistical methods we can prepare a large set of generalized string patterns from a raw corpus of natural texts. Then we can try to apply this material for the automatic construction of FOSC theories for specific applications. In this framework, the problem of generating a FOSC theory P can be formulated as follows.

Consider the following two sets of string pairs, which represent the desired and undesired input-output pairs respectively:

$E_+ = \{(s_i, t_i): i=1,2,\ldots, m\}$ *(Positive examples)*
$E_- = \{(s'_j, t'_j): j=1,2,\ldots, n\}$ *(Negative examples)*

A notation $Q(s)=t$ will be applied below for denoting a theory Q, which produces an output string t from an input string s in a limited time. A theory P can be automatically generated using the following objective function:

$$f(Q) = (|\{i: Q(s_i) = t_i\}| - |\{j: Q(s'_j) = t'_j\}|) / |Q|$$
$$P = \arg \max(f(Q))$$

where $|Q|$ denotes the total number of strings contained in the clauses of theory Q. The numerator of the objective function characterizes a quality of the theory Q regarding sets E_+ and E_-. The denominator characterizes a complexity of the theory Q. This objective function can specify how to choose among the many ways to structure and generalize a given set of strings and how to join the strings into clauses. Some modern optimization methods can be applied for the maximization of $f(Q)$.

We can use various sources of string samples for inclusion into a theory Q. For example, we can create such samples using a pair-wise alignment and most specific generalization of sentences, contained in a natural language corpus. Consider the following set of sentences (here we avoid capital letters so that they could be used for FOSC variables):

a book is in an open box
a notebook is in the closed box
the cat is in the cage
a fox is in a zoo
john is in his red car
this large table is red

A pair-wise alignment of these sentences provides a set of generalizations, which contains the following strings:

a [X] is in [D] [Y] box,
[X] is in [D] [Z],
a [X] is in [Z],
[X] is in [Z],
[X] is [A],

where variables X and Z stand for a noun or a noun group, D stands for determiners, Y stands for adjectives, and A stands for adjective phrases. The generalization is done by a substitution of the non-matching text constituents with variable symbols. Note that we get so generalized sentence patterns without any grammar or morphology models.

The discussable subject, suggested by these examples, is a possibility to construct a practical question-answering system using such material extracted from a large raw corpus of sentences. This section should be considered an invitation to carry out experiments in order to analyze this possibility for various problem domains, e.g., for developing a library-related dialogue system about book availability, or a system for a dialogue about geographical objects.

An important observation, supported by our examples, is that we don't need the explicitly defined qualifiers (formal grammar non-terminals): a noun-phrase, an adjective phrase etc. for the variables X, Z, A etc, although the variables may match only phrases of such types when the theories are applied to correct inputs.

For example, if a simple and correct English sentence is aligned with string "[X] is in [Z]", then we can expect that variables X and Z represent noun phrases, *although we do not explicitly define what a noun phrase is*. As a curious implication of this observation, we have a certain freedom from the ambiguity problems related to the usage of formal grammars. In fact, we don't apply formal grammars at all: FOSC provides the alternative options for text analysis and generation.

However, the multiple possibilities of alignment and unification of strings can impose their own ambiguity problems. For example, when string "[X] is in [Z]" is aligned with a complex raw sentence, the variables X and Z may represent not only noun phrases, but also some non-classifiable sentence fragments. We can approach to these problems using FOSC methods, without the application of external mechanisms for resolving ambiguity. For example, when aligning an English sentence with string "[X] is in [Z]", we can require that the text constituents for variables X and Z do not contain still non-structured substrings, which contain segment "is in".

Below we outline another direction for building FOSC theories. Here the composition of strings into clauses is controlled by a set E of ground string examples and a subset $G \subset G(E)$ of its generalization universe (see Definitions 2.5 and 2.10). Let us apply a notation $H \in \in G$ for stating that string H is a representative of some element of set G. Remind (Section 3) that a string theory is considered a *sequence* of clauses.

Algorithm Scheme 7.1. Build a string theory Q of a set E of ground string examples using generalizations $G \subset G(E)$.

1) *Take the most general element $H \in \in G$, not considered before, and put the clause H (with empty body) at the top of sequence Q. Stop stating an unsuccessful status if such element H does not exist.*
2) *If Q is a theory of set E, then remove all its clauses, which are not used in the application of Q to the elements of set E, and stop stating a successful status.*
3) *If Q contains a clause with empty body, find a bounding sequence B for its head H using a supporting example $\alpha \in E$. Proceed to Step 1 if this is impossible. Otherwise replace this clause with the description clause $H :- B$. Put into Q all the attractors of this bounding, which are still not the heads of clauses in Q, in the form of clauses with empty bodies. Proceed to Step 2.*

We justify our terminology of acceptors and attractors as follows. The heads of the clauses being created should *accept* the examples. The heads of already created clauses should *attract* the bounds when constructing the bounding sequences. In Step 3 we assume that the attractors and the bounding sequences are constructed from the representatives of elements of set G.

In order to apply this scheme, we should specify the following: 1) how a bounding sequence B is constructed; 2) in what positions of the theory P the new attractors are inserted; 3) how to determine whether a current theory is a theory of set E. The formation of set $G \subset G(E)$ should also be specified.

A possible strategy for the search of description clauses consists in the requirement that any description clause H :– B is supported by a *majority* of examples $\alpha \in E$, for which this clause is applicable. When sets E and G are sufficiently small, we can apply an exhaustive search among all possible bounding sequences and positions for inserting the attractors.

Example 7.1. Consider the ground string examples and four upper elements of the generalization universe described in Fig. 1 (Section 2). Assume that we have only three variables: A, B, C, and that we look only for clauses containing no more than one bound. In this case Algorithm Scheme 7.1 transforms into a simple exhaustive algorithm. The string theory of Example 2.1 is found by trying all possible bounding clauses and positions of the attractors.

In more complex cases we should seek for a partition of set E into simpler parts or for the application of special heuristics. In the tradition of ILP [18], the search for useful hypotheses can be combined with the application of background knowledge.

The search strategy of Algorithm Scheme 7.1 can be characterized as follows:

Try to apply the most general text pattern as a clause. If the pattern is too general, try to bind it into a description clause for some of the examples. If the theory still is not constructed, then put less general patterns on the top of the theory so that those too general become unattainable for the specific cases.

Generalization universe (cf. Section 2), derived from sentences of a natural language corpus, defines an ultimate expressive structure of the knowledge, represented in the corpus, without reference to any artificially formalized concepts. Doing a pairwise alignment and generalization of the sentences, contained in the corpus, we can expect that the frequency of the shorter generalizations is higher and that the shorter generalizations represent more correct, more general, and more applicable sentence patterns, like those applied in our examples in Section 4.

It is unrealistic and probably unnecessary to try to build the entire generalization universe G(E). For the natural language sentences, the set $G \subset G(E)$ can be limited in such a way that only the short and simple sentences are applied for the pair-wise generalization.

So, FOSC provides a universal framework of text perception, formalization, understanding and manipulation, which is independent of any prior formal knowledge. This framework allows us to immediately exploit the regularities and meanings, *implicitly* demonstrated in the text.

Here we see a philosophical challenge to the common usage of *explicit* methods, which often appear to be stiff, expensive, incomprehensible and mutually incompatible. Paradoxically, today the *implicit* methods are almost unknown, and most researchers define the semantics only explicitly, involving a large number of complex and hardly compatible formalisms resulted from the available models of various phenomena. In contrast to this, we try to develop a *universal* framework of intelligent text processing, which is independent of any prior knowledge.

There is another philosophical challenge, which is related to the usage of logic reasoning in general. Can we give up the logic connectives in the same way we have given up the logic predicates? In FOSC, we apply the traditional logic connectives for constructing and combining clauses. More specifically, we apply conjunction, negation, implication, and the resolution rule. So we manipulate with text semantics without predicate and formal grammar mechanisms. What will we gain if we remove the logical connectives as well? Can we model the text meaning using *only* the generalization, structurization and unification of text examples? Can we organize a *natural* reasoning, appealing only to the inherent logic of the text itself, without the first order logic inference? We finish this paper with this question unanswered, addressing it to a future research.

8   Related Work and the Ideas for Future Research

Modern information resources, such as Wikipedia, digital libraries, patent banks and technical regulations, contain a large amount of knowledge in the form of unstructured or semi-structured texts. In order to make inferences from this knowledge, attempts are made to structure it into a first order logic form. A common usage of logical methods assumes the design of logical predicates. This is done, for instance, in CycL [14] system. CycL contains a very large set of handcrafted first order logic micro-theories. However, this is a very expensive undertaking. Besides, the predicates reflect the view of their designer rather than the inherent logic of the subject.

We show that one can do first order logic without predicates at all, providing, in this way, a new direction for tackling this problem. The *implicit* definitions are much simpler to combine than the *explicit* definitions. Indeed, as in our speech, we can easily combine thousands of concepts without defining them explicitly. Furthermore, the way we apply the concepts assumes that they are already combined in some existing and correct text examples. In contrast, the attempts to combine *explicit* concepts typically require an expensive manual encoding and very complex axiomatization.

String unification enables us to implicitly operate with senses, related to the aligned text constituents. Unification of various kinds of structures (which have slots or variables for filling-in by the matching constituents) is applied in various fields. Unification of logical atoms is the most prominent example. Unification has been applied not only in a Logic Programming framework, as we do, and not only for strings or logical atoms.

Unification grammars (see, e.g., [10], [11]) present an example where a unification of special objects (feature structures) is applied in order to constrain the application of formal grammar rules. Note, however, that in this research the unifiable structures play a secondary role in reasoning. The primary role has the formal grammar. In a

Logic Programming framework the unification of objects has a more active role, supporting the implicit definition and manipulation of a larger class of senses.

Implicit formalization methods have their own history. In mathematical equations the implicit variables may denote functions, which do not have any explicit definition, or for which the explicit definitions are very complex and should be avoided. The following programming languages support the implicit definition of various senses by instantiation of variables through a matching of various structures. ANALYTIC language [7] contains special operators *compare* and *apply*, which are applied for the matching of symbolic structures containing variable symbols. For instance, the application of a symbolic form of differentiation rules to an analytic expression of a function produces an analytic expression of the function derivative. Matching of various structures, containing variables, is applied in Snobol [8], Refal [23] and Planner [9] languages. FOSC can be considered an attempt to extend these approaches for working with unrestricted strings.

ILP [18] suggests substantially developed inductive methods for the inference of predicate-logic programs from examples of logical atoms and a background knowledge, which is also expressed in a predicate-logic form. Grammar induction using ILP methods (LLL, Learning Language in Logic) is described in [5]; see also [4]. ILP methods can be similarly applied for the automatic extraction of FOSC theories from string examples. We consider this a prospective direction for future research. Some consideration of ILP-style methods for text-related inference is done in [6]. There is a certain analogy between a most specific common generalization of strings (Section 2) and a least general generalization of logical atoms, introduced by Plotkin [21].

An extensive analysis of various unification algorithms, applied in logical systems (including string unification algorithms), is given in [1]. In [13] an extension of classical first order logic with sequence variables and sequence functions is described. These constructions coexist with ordinary variables and functions of a fixed arity. A special syntax, semantics, inference system and inductive theory with sequence variables are developed in this calculus. G.S.Makanin [16] proved the computability of a word equation (string unification) problem in 1977. NP-hardness of this problem is shown in many publications, see e.g. [2]. We apply a limited, but efficient string unification algorithm, which is sufficient for achieving Turing completeness. Various methods of string alignment are developed and applied in biotechnology and NLP.

The incorporation of common-sense reasoning into computer programs is currently an active research area. For instance, an attempt of this kind, related to building a large ontology, is done in SUMO [19] project. Our method of implicit definition of senses is another attempt to apply existing common-sense reasoning patterns in computer programs.

An *automatic* extraction of FOSC theories from texts, contained in the information resources mentioned above, is a perspective direction for the future development of the method. Thesis 3.1 suggests that, given a sufficiently rich set E of string examples which *demonstrate* a theory or algorithm, we can always build this theory or algorithm from the generalizations of the examples. The generalization universe G(E) (cf. Section 2) provides all possible building blocks for such a theory.

So, the direction for a future research, which we suggest here, is finding a way of extracting and composing such blocks into a theory automatically, without inventing *artificial* notations for the designation of the involved concepts and relations.

This direction will hopefully provide new insights for the analysis of biological sequences: since we apply the unrestricted string patterns immediately in reasoning, we may compose a FOSC theory, which employs the inherent meaning of such sequences and does not depend on any previous formal knowledge.

9 Conclusion

First Order String Calculus (FOSC) is an attempt to build a cognitive model of the brain by simulating the matching, generalization, structurization and composition of the images contained in our memory. FOSC can be applied for the extraction of formal theories and algorithms immediately from unrestricted string examples, without any previous semantic knowledge, expressed mathematically. The regularities, demonstrated in similar text examples, are extracted using an alignment and a generalization of the examples. The algorithmic forms of these regularities – string generalizations, structurizations and compositions into clauses – simultaneously serve as elements of the theory of the examples and as a universal algorithm definition means. This way of creating algorithms is Turing-complete and provides a reasoning framework, which is at least as powerful as a predicate logic, based on the axiomatic principles. Furthermore, FOPC is a special case of FOSC.

FOSC theories provide an alternative to both predicate-logic-based and formal-grammar-based approaches to text analysis and generation.

The method of implicit semantics, based on string alignment, is an adaptation of the traditional method of implicit function definition in mathematical equations, for the fields of logics, linguistics, and intelligent text processing in general. Like in traditional mathematics, this method allows one to apply the inherent meaning of the expression constituents and avoid the complex formulas, which are necessary when the meaning is defined explicitly.

FOSC, along with the generalization universe of a set of unrestricted strings, can be considered a kind of concept algebra, where the concepts are defined implicitly. The generalization universe can be applied in order to assess the expressive power and cognitive limits of the knowledge, expressed in the unrestricted texts, independently of any other knowledge.

References

[1] Baader, F., Snyder, W.: Unification Theory. Handbook of Automated Deduction, ch. 8. Springer, Berlin (2001)

[2] Černál, I., Klíma, O., Srba1, J.: On the Pattern Equations. FI MU Report Series, FIMU-RS-99-01 (1999)

[3] Church, A.: Introduction to Mathematical Logic, vol. 1. Princeton University Press, Princeton (1956)

[4] Cicchello, O., Kremer, S.C.: Inducing Grammars from Sparse Data Sets: A survey of Algorithms and Results. Journal of Machine Learning Research 4, 603–632 (2003)

[5] Dzeroski, S., Cussens, J., Manandhar, S.: An Introduction to Inductive Logic Programming and Learning Language in Logic. In: Cussens, J., Džeroski, S. (eds.) LLL 1999. LNCS, vol. 1925, pp. 4–35. Springer, Heidelberg (2000)

[6] Gleibman, A.H.: Knowledge Representation via Verbal Description Generalization: Alternative Programming in Sampletalk Language. In: Workshop on Inference for Textual Question Answering, July 2009, 2005 – Pittsburgh, Pennsylvania, pp. 59–68, AAAI 2005 - the Twentieth National Conference on Artificial Intelligence, http://www.hlt.utdallas.edu/workshop2005/papers/WS505GleibmanA.pdf

[7] Glushkov, V.M., Grinchenko, T.A., Dorodnitcina, A.A.: Algorithmic Language ANA-LYTIC-74. Kiev. Inst. of Cybernetics of the Ukraine Academy of Sciences (1977) (in Russian)

[8] Griswold, R.E.: The Macro Implementation of SNOBOL4. W.H. Freeman and Company, San Francisco (1972)

[9] Hewitt, C.E.: Description and theoretical analysis (using schemata) of PLANNER: a language for proving theorems and manipulating models in a robot. Technical Report, AI-TR-258, MIT Artificial Intelligence Laboratory (1972)

[10] Jaeger, E., Francez, N., Wintner, S.: Unification Grammars and Off-Line Parsability. Journal of Logic, Language and Information 14, 234–299 (2005)

[11] Jurafsky, D., Martin, J.H.: Speech and Language Processing. An introduction to Natural Language Processing, Computational Linguistics, and Speech Recognition. Prentice-Hall, Englewood Cliffs (2000)

[12] Kleene, S.C.: Mathematical Logic. John Wiley & Sons, Chichester (1967)

[13] Kutsia, T., Buchberger, B.: Predicate Logic with Sequence Variables and Sequence Function Symbols. In: Asperti, A., Bancerek, G., Trybulec, A. (eds.) MKM 2004. LNCS, vol. 3119, pp. 205–219. Springer, Heidelberg (2004)

[14] Lenat, D.B.: From 2001 to 2001: Common Sense and the Mind of HAL. In: Stork, D.G. (ed.) HAL's Legacy: 2001's Computer as Dream and Reality. MIT Press, Cambridge (2002)

[15] Lloyd, J.W.: Foundations of logic programming. Artificial Intelligence Series. Springer, New York (1987)

[16] Makanin, G.S.: The Problem of Solvability of Equations in a Free Semigroup. Mat. Sbornik. 103(2), 147–236 (in Russian); English translation in: Math. USSR Sbornik 32, 129–198 (1977)

[17] Markov, A.A.: Theory of Algorithms. Trudy Mathematicheskogo Instituta Imeni V. A. Steklova 42 (1954) (in Russian)

[18] Muggleton, S.H., De Raedt, L.: Inductive Logic Programming: Theory and Methods. Logic Programming 19(20), 629–679 (1994)

[19] Niles, I., Pease, A.: Towards a Standard Upper Ontology. In: Welty, C., Smith, B. (eds.) Proceedings of the 2nd International Conference on Formal Ontology in Information Systems (FOIS 2001), Ogunquit, Maine (2001)

[20] Piaget, J., Inhelder, B., Weaver, H.: The Psychology of the Child. Basic Books (1969)

[21] Plotkin, G.: A note on inductive generalization. Machine Intelligence, vol. 5, pp. 153–163. Edinburgh University Press (1970)

[22] Robinson, J.A.: A Machine-oriented Logic Based on the Resolution Principle. J. ACM 12(1), 23–41 (1965)

[23] Turchin, V.F.: Basic Refal. Language Description and Basic Programming Methods (Methodic Recommendations), Moscow, CNIIPIASS (1974) (in Russian)

[24] Vigandt, I.: Natural Language Processing by Examples. M. Sci. Thesis, Comp. Sci. Dept., Technion, Haifa, 115 p. (1997) (in Hebrew, with abstract in English)

Appendix: The Applied String Unification Algorithm

In this appendix we describe a string unification algorithm, which is applied in our implementation of Sampletalk compiler [6]. Let (S_1, S_2) be a pair of strings, which can be represented by any of the following alignments (here the indexes define the correspondence between the aligned segments):

$$S_1 = s_1 V s_3$$
$$S_2 = s_1 g_2 g_3 \tag{1}$$
$$S_1 = s_1 s_2 s_3$$
$$S_2 = s_1 W g_3 \tag{2}$$

Here V and W are variables, which do not occur in segments g_2 and s_2 respectively, and segments s_3 and g_3 are either empty or start with compatible symbols, which means that either $s_3(0) = g_3(0)$ or at least one of these symbols is a variable. (We apply a 0-based indexation of the symbols in strings). Segment s_1 also may be empty. Segment g_2 (s_2) is assumed to be a valid string (concerning the nesting structure, defined by the square bracket pairs) and also may start from a variable symbol. Variables V and W are called *bound* by segments g_2 and s_2 respectively.

Such an alignment is called *a leftmost bounding hypothesis* for string pair (S_1, S_2). There can be more than one leftmost bounding hypothesis for the same string pair. As an example, consider two alignments of string "[A , B] $\Rightarrow$ [C , A]" with string "[c a t , d o g , m o n k e y] $\Rightarrow$ [W]", described in Section 2, where variable A is bound by segment "c a t" in one case and by segment "c a t , d o g" in another.

Given a string pair (S_1, S_2), we consider all leftmost bounding hypotheses sequentially according to the following convention:

a) If $g_2(0)$ is a terminal symbol (case 1), then the versions for g_2 are ordered by incrementing the number of symbols in segment g_2;

b) If $s_2(0)$ is a terminal symbol (case 2), then the versions for s_2 are ordered by incrementing the number of symbols in segment s_2;

c) Otherwise (a combined case where both $g_2(0)$ and $s_2(0)$ are variables) we consider all versions for g_2, organized similarly to the case (a), followed by all versions for s_2, organized similarly to the case (b).

Since the number of symbols in a string is finite, the sequence of leftmost bounding hypotheses is finite for any string pair (S_1, S_2). The following recursive procedure sequentially produces the unifications of string pair (S_1, S_2):

Procedure 1. (String Unification Algorithm): Find unifications of a string pair (S_1, S_2).

1) *If $S_1 = S_2$ then report string S_1 and exit.*
2) *Find a leftmost bounding hypothesis, not considered before, for the string pair (S_1, S_2). If such hypothesis does not exist then exit.*
3) *Replace all occurrences of the bound variable in string pair (S_1, S_2) with the corresponding bounding segment.*
4) *Apply Procedure 1 to the string pair (S_1, S_2), modified in Step 3.*
5) *Undo the change made in Step 3 and proceed to Step 2.*

We assume that, along with reporting every unification string, this procedure reports the corresponding substitution of variables, which is applied for the original string pair (S_1, S_2) in order to produce this unification string.

We should note that the presented algorithm has certain limitations. Particularly, we cannot be sure that the algorithm produces *all* unifications of a given pair of string. In spite of these limitations, the algorithm can be applied in a Turing-complete algorithm definition system (see Theorem 3.1).

String unification algorithm, defined above, organizes all the possibilities to unify a pair of strings in a specific order, which we call *a canonical order*. The unifications are ordered and can be sequentially applied according to this order. All possible instantiations of a clause H :– B by a literal S (if there are many) are also ordered according to the canonical order of possible unifications of strings S and H. This order is used in the application of a string theory to a string (Algorithm 3.1).

Knowledge Reduction in Concept Lattices Based on Irreducible Elements

Xia Wang[1],[*] and Wenxiu Zhang[2]

[1] Department of Applied Mathematics and Physics, Institute of Science
PLA University of Science and Technology, Nanjing, Jiangsu, 211101, China
[2] Institute for Information and System Sciences, Faculty of Science
Xi'an Jiaotong University, Xi'an, Shaan'xi, 710049, China
bblylm@126.com

Abstract. As one of the important problems of knowledge discovery and data analysis, knowledge reduction can make the discovery of implicit knowledge in data easier and the representation simpler. In this paper, a new approach to knowledge reduction in concept lattices is developed based on irreducible elements, and characteristics of attributes and objects are also analyzed. Furthermore, algorithms for finding attribute and object reducts are provided respectively. The algorithm analysis shows that the approach to knowledge reduction involves less computation and is more tractable compared with the current methods.

Keywords: Concept lattice, knowledge reduction, irreducible element, attribute characteristic.

1 Introduction

Formal concept analysis [1,2] is an effective tool for knowledge representation and knowledge discovery. The basic notions of formal concept analysis are concepts and concept lattices. A concept lattice is an ordered hierarchical structure of concepts which are defined by a binary relation between a set of objects and a set of attributes. Although rough set theory [3] and formal concept analysis are two different theories, they have much in common in terms of both goals and methodologies [4], and relationship between formal concept analysis and rough set theory is studied vastly [4,5,6,7,8,9,10]. Other researches on concept lattices focus on such topics as [11]: construction of concept lattices [12,13,14], pruning of concept lattices [15], acquisition of rules [13,14], and applications [15,16].

In recent years, much attention has been devoted to knowledge reduction in formal concept analysis. An approach to attribute reduction in concept lattices has been presented based on discernibility matrix [17,18,19]. The approach preserves all of the concepts and their hierarchy in a context, however, it needs to generate all concepts and compute the discernibility matrix of the context. The algorithms for generating all concepts of a context mainly include: the Naive

---

[*] Corresponding author.

M.L. Gavrilova et al. (Eds.): Trans. on Comput. Sci. V, LNCS 5540, pp. 128–142, 2009.

algorithm, the next closure algorithm, and the object intersections algorithm [20], and it is shown that the algorithms for generating all concepts involve fussy computation. In [21], attribute reduction in concept lattices is also studied from the viewpoint of dependence space of contexts, but the complicated calculations are still its shortcoming. Therefore, it is required to develop some more effective methods for knowledge reduction in concept lattices. In [22,23], a new idea of attribute reduction in concept lattices, attribute oriented concept lattices and object oriented concept lattices has been developed by means of $\wedge-$irreducible elements. This kind of method needs not generate all concepts and sounds simple to compute attribute reducts. However, there are no explicit explains or specific algorithms to illustrate that.

In this paper, properties of irreducible elements of a context are studied in detail, and relationships of irreducible elements between the original context and its subcontext are also obtained. Then based on irreducible elements, a new approach to knowledge reduction including attribute reduction and object reduction in concept lattices is presented. The main idea of this approach to attribute (object) reduction is to find minimal attribute (object) subsets, which preserve the extent (intent) set of irreducible elements of the original context. It is further proved that the attribute (object) reduct also preserves all concepts and their hierarchy. Moreover, algorithms for finding attribute reducts and object reducts are given and analyzed respectively. Compared with the current methods for knowledge reduction, the method provided in this paper involves less computation and is more tractable.

This paper is organized as follows. In Sect. 2, we recall basic definitions of formal concept analysis. In Sect. 3, relationships of irreducible elements between the original context and its subcontext are studied in detail. In Sect. 4, definitions and main results of the developed approach to knowledge reduction including attribute reduction and object reduction in concept lattices are provided. And algorithms of knowledge reduction are also presented and analyzed. Finally, we conclude the paper in Sect. 5.

2   Basic Definitions of Formal Concept Analysis

To facilitate our discussions, some basic notions and properties about formal concept analysis are introduced in this section which can be found in [2].

Definition 1. *A (formal) context (U, A, I) consists of two sets U and A, and a relation I between U and A. The elements of U are called the objects and the elements of A are called the attributes of the context.*

For every $X \subseteq U$ and $B \subseteq A$, Ganter and Wille defined two operators,

$$X^* = \{a \in A | \forall x \in X, (x, a) \in I\}, \tag{1}$$

$$B^* = \{x \in U | \forall a \in B, (x, a) \in I\}. \tag{2}$$

For the sake of simplicity, we write x^* instead of $\{x\}^*$ and a^* instead of $\{a\}^*$ for all $x \in U$, $a \in A$. A context is called canonical, if $x^* \neq \emptyset$, $x^* \neq A$ for all

$x \in U$, and $a^* \neq \emptyset$, $a^* \neq U$ for all $a \in A$. The contexts in the following are canonical.

Definition 2. *A (formal) concept of a context (U, A, I) is a pair (X, B) with $X \subseteq U$, $B \subseteq A$, $X^* = B$ and $B^* = X$. We call X the extent and B the intent of the concept (X, B).*

The concepts of a context (U, A, I) are ordered by

$$(X_1, \ B_1) \leq (X_2, \ B_2) \Leftrightarrow X_1 \subseteq X_2 (\Leftrightarrow B_1 \supseteq B_2),$$

where $(X_1, \ B_1)$ and $(X_2, \ B_2)$ are two concepts. The set of all concepts of (U, A, I) ordered in this way is denoted by $L(U, A, I)$ and is called the concept lattice of the context (U, A, I). The infimum and supremum are given by

$$\bigwedge_{t \in T} (X_t, B_t) = (\bigcap_{t \in T} X_t, (\bigcup_{t \in T} B_t)^{**}), \tag{3}$$

$$\bigvee_{t \in T} (X_t, B_t) = ((\bigcup_{t \in T} X_t)^{**}, \bigcap_{t \in T} B_t). \tag{4}$$

Then the concept lattice $L(U, A, I)$ is a complete lattice, and the pair of functions $*$ induces a Galois connection between $\mathcal{P}(U)$ (where $\mathcal{P}(X)$ denotes the power set of X) and $\mathcal{P}(A)$. The two functions $* : \mathcal{P}(U) \to \mathcal{P}(A)$ and $* : \mathcal{P}(A) \to \mathcal{P}(U)$ satisfy the following properties:

Proposition 1. *Let (U, A, I) be a context, X, X_t be sets of objects, and B, B_t be sets of attributes for every $t \in T$. Then,*

(1) $X_1 \subseteq X_2 \Rightarrow X_2^* \subseteq X_1^*$, (1^∂) $B_1 \subseteq B_2 \Rightarrow B_2^* \subseteq B_1^*$,
(2) $X \subseteq X^{**}$, (2^∂) $B \subseteq B^{**}$,
(3) $X^* = X^{***}$, (3^∂) $B^* = B^{***}$,
(4) $(\bigcup_{t \in T} X_t)^* = \bigcap_{t \in T} X_t^*$, (4^∂) $(\bigcup_{t \in T} B_t)^* = \bigcap_{t \in T} B_t^*$.

By Proposition 1(3), for any object set $X \subseteq U$, (X^{**}, X^*) is a concept. Similarly, (B^*, B^{**}) is also a concept for any attribute set $B \subseteq A$. In particular, (x^{**}, x^*) and (a^*, a^{**}) are called the object concept and the attribute concept respectively for all $x \in U$ and $a \in A$.

We call $L_U(U, A, I) = \{X | (X, B) \in L(U, A, I)\}$ the extent set, and $L_A(U, A, I) = \{B | (X, B) \in L(U, A, I)\}$ the intent set of the context (U, A, I).

Definition 3. *Let $T = (U, A, I)$ be a context. For any $D \subseteq A$, the context $T_D = (U, D, I_D)$ is called a subcontext of T, where $I_D = I \cap (U \times D)$ and $X^{*D} = X^* \cap D$.*

Clearly, for any $D \subseteq A$, $L_U(T_D) \subseteq L_U(T)$.

3   Irreducible Elements

Proposition 13[2] gives a method for judging the irreducible element using an arrow relation which is defined between a single attribute and a single object. In this section, we study properties of irreducible elements by the extents of attribute concepts and the intents of object concepts.

Definition 4. *[24]   Let L be a lattice. An element $a \in L$ is $\wedge$-irreducible if (1)$a \neq 0$ (in case L has a zero), (2)$a = b \wedge c$ implies $a = b$ or $a = c$ for any $b, c \in L$. Dually, an element $a \in L$ is $\vee$-irreducible element if (1)$a \neq 1$ (in case L has a unit), (2)$a = b \vee c$ implies $a = b$ or $a = c$ for any $b, c \in L$.*

Theorem 1. *Let L be a finite lattice. Every element in L is a join (meet, resp.) of some $\wedge$-irreducible ($\vee$-irreducible, resp.) elements.*

Let $T = (U, A, I)$ be a context. Since for any $(X, B) \in L(T)$

$$(X, B) = \bigwedge_{a \in B} (a^*, a^{**}), \quad (X, B) = \bigvee_{x \in X} (x^{**}, x^*), \tag{5}$$

we have that a $\wedge$-irreducible element must be an attribute concept, and a $\vee$-irreducible element must be an object concept. We express the $\wedge$-irreducible element and $\vee$-irreducible element by the extents of attribute concepts and the intents of object concepts respectively.

Lemma 1.   *Let $T = (U, A, I)$ be a context. $\forall a \in A$ and $\forall x \in U$, we have*

(i) (a^*, a^{**}) is $\wedge$-irreducible $\Leftrightarrow \{g \in A | a^* \subsetneq g^*\} = \emptyset$, or $\{g \in A | a^* \subsetneq g^*\} \neq \emptyset$ implies $a^* \neq \bigcap_{a^* \subsetneq g^*} g^*$.

(ii) (x^{**}, x^*) is $\vee$-irreducible $\Leftrightarrow \{y \in U | x^* \subsetneq y^*\} = \emptyset$, or $\{y \in U | x^* \subsetneq y^*\} \neq \emptyset$ implies $x^* \neq \bigcap_{x^* \subsetneq y^*} y^*$.

Lemma 1 can be proved directly by Proposition 13[2], and it shows that an attribute concept is $\wedge$-irreducible if and only if it is the maximal element of attribute concepts, or its extent can not be represented as the meet of extents of attribute concepts which properly include its extent. Therefore, we can obtain all $\wedge$-irreducible elements just from the extent set of attribute concepts $\{a^* | a \in A\}$. Dually, all $\vee$-irreducible elements can be judged from the intent set of object concepts $\{x^* | x \in U\}$.

Let $T = (U, A, I)$ be a context. $\forall D \subseteq A, \forall V \subseteq U, M(L(T_D))$ and $J(L(T_V))$ represent the set of $\wedge$-irreducible elements in $L(T_D)$ and the set of $\vee$-irreducible elements in $L(T_V)$ respectively. And we call $M_U(L(T_D)) = \{X | (X, B) \in M(L(T_D))\}$ the extent set of $\wedge$-irreducible elements in $L(T_D)$ and $J_A(L(T_V)) = \{B | (X, B) \in J(L(T_V))\}$ the intent set of $\vee$-irreducible elements in $L(T_V)$.

The following three lemmas describe the relationships of irreducible elements between an original context and its subcontext.

Lemma 2. *Let $T = (U, A, I)$ be a context and D a subset of A. For all $d \in D$, if $d^* \in M_U(L(T))$, then $d^* \in M_U(L(T_D))$.*

Proof. Readily, $\{g \in D | d^* \subsetneq g^*\} \subseteq \{g \in A | d^* \subsetneq g^*\}$. Now we prove the result from three cases: (i) $\{g \in A | d^* \subsetneq g^*\} = \emptyset$, (ii) $\{g \in D | d^* \subsetneq g^*\} \neq \emptyset$, and (iii)$\{g \in D | d^* \subsetneq g^*\} = \emptyset$ and $\{g \in A | d^* \subsetneq g^*\} \neq \emptyset$.

(i) Clearly, $\{g \in A | d^* \subsetneq g^*\} = \emptyset$ implies $\{g \in D | d^* \subsetneq g^*\} = \emptyset$. Therefore, if $\{g \in A | d^* \subsetneq g^*\} = \emptyset$, then $d^* \in M_U(L(T))$ implies $d^* \in M_U(L(T_D))$ by Lemma 1(i).

(ii) If $\{g \in D | d^* \subsetneq g^*\} \neq \emptyset$, then $\{g \in A | d^* \subsetneq g^*\} \neq \emptyset$. Because $d^* \subseteq \bigcap\limits_{d^* \subsetneq g^*, g \in A} g^* \subseteq \bigcap\limits_{d^* \subsetneq g^*, g \in D} g^*$ holds, if $d^* \neq \bigcap\limits_{d^* \subsetneq g^*, g \in A} g^*$ then $d^* \neq \bigcap\limits_{d^* \subsetneq g^*, g \in D} g^*$.
Combined with Lemma 1(i), if $\{g \in D | d^* \subsetneq g^*\} \neq \emptyset$, then $d^* \in M_U(L(T))$ implies $d^* \in M_U(L(T_D))$.

(iii) If $\{g \in D | d^* \subsetneq g^*\} = \emptyset$ and $\{g \in A | d^* \subsetneq g^*\} \neq \emptyset$, then $d^* \in M_U(L(T_D))$ according to Lemma 1(i). Obviously, if $d^* \in M_U(L(T))$, then $d^* \in M_U(L(T_D))$.

Therefore, the result is concluded.   □

It is obvious that for all $a \in A$, if $a^* \in M_U(L(T))$ and there exists $d \in D$ such that $a^* = d^*$, then $M_U(L(T)) \subseteq M_U(L(T_D))$ due to Lemma 2. The following lemma gives the condition under which $M_U(L(T)) = M_U(L(T_D))$.

Lemma 3. *Let $T = (U, A, I)$ be a context. $\forall D \subseteq A$, we have*

$$M_U(L(T)) \subseteq M_U(L(T_D)) \Leftrightarrow M_U(L(T)) = M_U(L(T_D)).$$

Proof. Necessity. Assume $M_U(L(T)) \neq M_U(L(T_D))$. Then there exists $a \in A$ such that $a^* \in M_U(L(T_D))$ and $a^* \notin M_U(L(T))$. Thus, $\{g \in A | a^* \subsetneq g^*\} \neq \emptyset$ and $a^* = \bigcap\limits_{a^* \subsetneq g^*, g \in A} g^*$ hold by Lemma 1(i), and $a^* = \bigcap\limits_{\substack{a^* \subsetneq h^*, \\ h^* \in M_U(L(T))}} h^*$ holds by Theorem 1. Combined with $M_U(L(T)) \subseteq M_U(L(T_D))$, $a^* \notin M_U(L(T_D))$ due to Lemma 1(i), which is a contradiction to $a^* \in M_U(L(T_D))$. Therefore, $M_U(L(T)) = M_U(L(T_D))$.
Sufficiency. It is trivially.   □

The following lemma follows directly from Lemma 2 and Lemma 3:

Lemma 4. *Let $T = (U, A, I)$ be a context. For any $D \subseteq A$, we have $M_U(L(T)) = M_U(L(T_D)) \Leftrightarrow$ there exists $G \subseteq A$ such that $M_U(L(T)) = \{a^* | a \in G\}$ and $G \subseteq D$.*

From Lemma 4, we know that the attribute sets corresponding to $\wedge$-irreducible elements in a concept lattice preserve the extents of $\wedge$-irreducible elements in the original concept lattice. At the same time Lemma 4 gives a method for finding attribute subsets which preserve the extent set of $\wedge$-irreducible elements in the original concept lattice. The similar result of $\vee$-irreducible elements can be easily obtained.

Lemma 5. *Let $T = (U, A, I)$ be a context. For any $V \subseteq U$, the following statements hold:*

(1) $\forall x \in V$, $x^* \in J_A(L(T))$ *implies* $x^* \in J_A(L(T_V))$,

(2) $J_A(L(T)) \subseteq J_A(L(T_V)) \Leftrightarrow J_A(L(T)) = J_A(L(T_V))$.

(3) $J_A(L(T)) = J_A(L(T_V)) \Leftrightarrow$ *there exists* $W \subseteq U$ *such that* $J_A(L(T)) = \{x^* | x \in W\}$ *and* $W \subseteq V$.

Example 1. Table 1 gives a context $T = (U, A, I)$ with $U = \{1, 2, 3, 4, 5, 6\}$ and $A = \{a, b, c, d, e, f\}$. For simplicity, we omit the bracket pairs of attribute set and object set. For example, $a^* = 1256$ stands for $a^* = \{1, 2, 5, 6\}$. Then, the extents of attribute concepts and the intents of object concepts of the context T are

Table 1. A context $T = (U, A, I)$

U	a	b	c	d	e	f
1	1	0	1	1	1	0
2	1	0	1	0	0	0
3	0	1	0	0	1	1
4	0	1	0	0	1	1
5	1	0	0	0	0	0
6	1	1	0	0	1	1

$a^* = 1256$, $b^* = f^* = 346$, $c^* = 12$, $d^* = 1$, $e^* = 1346$, and
$1^* = acde$, $2^* = ac$, $3^* = 4^* = bef$, $5^* = a$, $6^* = abef$.

It is easy to verify that $d^* = c^* \cap e^*$, and each extent of attribute concepts except d^* can not be represented as the meet of extents which properly include it. Consequently, all attribute concepts except (d^*, d^{**}) are $\wedge$-irreducible by Lemma 1. In the same way, we can obtain that all object concepts except $(5^{**}, 5^*)$ are $\vee$-irreducible.

Therefore, the extent set of $\wedge$-irreducible elements and the intent set of $\vee$-irreducible elements are

$$M_U(L(T)) = \{a^*, b^*, c^*, e^*, f^*\} = \{a^*, b^*, c^*, e^*\} = \{a^*, c^*, e^*, f^*\}$$
$$= \{1256, 346, 12, 1346\}$$
$$J_A(L(T)) = \{1^*, 2^*, 3^*, 4^*, 6^*\} = \{1^*, 2^*, 3^*, 6^*\} = \{1^*, 2^*, 4^*, 6^*\}$$
$$= \{acde, ac, bef, abef\}.$$

Let $D_1 = \{a, b, c, e, f\}, D_2 = \{a, b, c, e\}, D_3 = \{a, c, e, f\}, V_1 = \{1, 2, 3, 4, 6\}$, $V_2 = \{1, 2, 3, 6\}$, and $V_3 = \{1, 2, 4, 6\}$. Clearly, $D_i, i = 1, 2, 3$ are the corresponding attribute subsets of $\wedge$-irreducible elements, and $V_j, j = 1, 2, 3$ are the corresponding object subsets of $\vee$-irreducible elements. According to Lemma 4 and Lemma 5, $M_U(L(T_{D_i})) = M_U(L(T))$ and $J_A(L(T_{V_j})) = J_A(L(T)), i, j = 1, 2, 3$.

Obviously, $D_i \subsetneq D_1, i = 2,3$, and for any $D \subsetneq D_1$, if $D \neq D_i, i = 2,3$, then $D \subsetneq D_2$ or $D \subsetneq D_3$. Similarly, $V_j \subsetneq V_1, j = 2,3$, and for any $V \subsetneq V_1$, if $V \neq V_j, j = 2,3$, then $V \subsetneq V_2$ or $V \subsetneq V_3$. Furthermore, it is easy to see that for any $D \subsetneq A$ if $D \subsetneq D_2$ or $D \subsetneq D_3$, then $M_U(L(T_D)) \neq M_U(L(T))$, and for any $V \subsetneq U$ if $V \subsetneq V_2$ or $V \subsetneq V_3$, then $J_A(L(T_V)) \neq J_A(L(T))$.

4   Knowledge Reduction in Concept Lattices Based on Irreducible Elements

Considering the important role of the extent set of irreducible elements played, we present a new approach to knowledge reduction, and study the characteristics of attributes and objects based on irreducible elements in this section.

4.1   Attribute Reduction in Concept Lattices

Wang and Ma [22] presented a definition of attribute reduction in concept lattices based on ∧-irreducible elements. The main idea of the definition is to find minimal attribute subsets which can preserve the extent set of ∧-irreducible elements in the original concept lattice. The following definition can be found in [22]:

Definition 5. *Let $T = (U, A, I)$ be a context. An attribute subset $D \subseteq A$ is called an attribute consistent set of T if D satisfies*

$$M_U(L(T)) = M_U(L(T_D)). \tag{6}$$

If D is an attribute consistent set of T, and there is no proper subset E of D such that E is a consistent set of T, then D is called an attribute reduct of T.

Theorem 2. *Let $T = (U, A, I)$ be a context. For any $D \subseteq A$, we have*

(i) D is an attribute consistent set of $T \Leftrightarrow L_U(T_D) = L_U(T)$,
(ii) D is an attribute reduct of $T \Leftrightarrow L_U(T_D) = L_U(T)$, and $L_U(T_{D-\{d\}}) \neq L_U(T)$ for all $d \in D$.

Proof

(i) Sufficiency. It is trivially.
     Necessity. It is obvious $L_U(T_D) \subseteq L_U(T)$ for any $D \subseteq A$. Now we only need to prove $L_U(T) \subseteq L_U(T_D)$. For any $X \in L_U(T)$, there exist $X_i \in M_U(L(T)), i \leq l$ such that $X = \bigcap_{i=1}^{l} X_i$ due to Theorem 1. Thus, $M_U(L(T)) = M_U(L(T_D))$ implies $X_i \in M_U(L(T_D))$ for each $i \leq l$. Since $M_U(L(T_D)) \subseteq L_U(L(T_D))$, we have $\bigcap_{i=1}^{l} X_i \in L_U(T_D)$, that is, $X \in L_U(T_D)$. Therefore, $L_U(T) \subseteq L_U(T_D)$.
(ii) The result follows directly from (i) and Definition 5.   □

Theorem 2 shows that an attribute consistent set preserves the extent set of the original concept lattice. Since $L_U(U, A, I)$ and $L_A(U, A, I)$ are dual and isomorphic, we can easily obtain that for any contexts (U_1, A_1, I_1) and (U_2, A_2, I_2) if $L_U(U_1, A_1, I_1) = L_U(U_2, A_2, I_2)$, then the two concept lattices $L(U_1, A_1, I_1)$ and $L(U_2, A_2, I_2)$ are isomorphic; dually if $L_A(U_1, A_1, I_1) = L_A(U_2, A_2, I_2)$, then $L(U_1, A_1, I_1)$ and $L(U_2, A_2, I_2)$ are also isomorphic. Further, we have that the concept lattice determined by an attribute reduct is isomorphic to the original concept lattice. Therefore, the method produced by Definition 5 is equivalent to that in [17,18,19]. Now we compare the two methods to illustrate that our method is more effective.

An algorithm for finding an attribute reduct using the method provided by Definition 5 is given as follows.

Input: (U, A, I)
Output: an attribute reduct D
1. $E = A$
2. $D = \emptyset$
3. **for** $(i = 1; i \leq n; i++)$ $\qquad\qquad (n = |A|)$
4. $\qquad \{a_i^* = \emptyset;$
5. $\qquad$ **for** $(j = 1; j \leq m; j++)$ $\qquad (m = |U|)$
6. $\qquad\quad \{$**if** $((x_j, a_i) \in I)\ a_i^* = a_i^* \cup \{x_j\};\}$
7. $\qquad \}$
8. **for** $(i = 1; i < n; i++)$
9. $\qquad \{$ **if** $(a_i \notin E)$ continue;
10. $\qquad$ **for** $(j = i + 1; j \leq n; j++)$
11. $\qquad\quad \{$ **if** $(a_j \notin E)$ continue;
12. $\qquad\qquad$ **if** $(a_i^* = a_j^*)\ E = E - \{a_j\};$
13. $\qquad\quad \}$
14. $\qquad \}$
15. Let $\{e_1, e_2, ..., e_{|E|}\} = E;$
16. **for** $(i = 1; i \leq |E|; i++)$
17. $\qquad \{G_i = \emptyset;$
18. $\qquad$ **for** $(j = 1; j \leq |E|; j++)$
19. $\qquad\quad \{$**if** $(e_i^* \subseteq e_j^*\ \&\&\ e_i^* \neq e_j^*)\ G_i = G_i \cup \{e_j\};\}$
20. $\qquad$ **if** $(G_i \neq \emptyset\ \&\&\ e_i^* = \bigcap_{e_j \in G_i} e_j^*)\ D = D;$
21. $\qquad$ **else** $D = D \cup \{e_i\};$
22. $\qquad \}$

Then D is an attribute reduct of the context (U, A, I), where $|X|$ is the cardinality of X. The worst case time complexity of the algorithm is as follows. Lines 3 and 5 can be done in $O(n \cdot m)$ (i.e. $O(|A| \cdot |U|)$). As lines 8 and 10 are repeated at almost $(n - 1) \cdot |E|$ times, their contribution to the final complexity is $O(n \cdot |E|)$ (i.e. $O(|A| \cdot |E|)$). Lines 16 and 18 can be computed in $O(|E|^2)$. Therefore, if $|E| \leq |U|$, then the final complexity is $O(|A| \cdot |U|)$; if $|E| > |U|$, then the final complexity is $O(|A| \cdot |E|)$.

In what follows, we list the main steps of computing attribute reducts by the method presented in [17,18,19]:

Step 1. construct the set of all concepts of the context (U, A, I);
Step 2. generate the discernibility matrix of (U, A, I);
Step 3. find the minimal attribute subset D such that the intersection of D and each nonempty discernibility attribute set is not empty.

The methods for constructing all concepts mainly include the Naive algorithm, the next closure algorithm, and the object intersections algorithm [20]. The Naive algorithm is extremely inefficient as it requires the consideration of all the subsets of U or A together with their corresponding closures, regardless of the size of the concept lattice being generated. And the last two algorithms are relatively more effective: the worst case time complexity of the next closure algorithm is $O(2^{|U|} \cdot |A| \cdot |U|)$, and that of the object intersections algorithm is $O(|A| \cdot |U| \cdot |C|)$ where $|C|$ is the number of the concepts.

Since $|C| \geq |E|$, we have $|A| \cdot |U| \cdot |C| > |A| \cdot |E|$, and clearly $|A| \cdot |U| \cdot |C| > |A| \cdot |U|$ holds. Therefore, the worst case time complexity of the algorithm for constructing all concepts is higher than that of our algorithm for finding an attribute reduct. Consequently, the method proposed by Definition 5 is more tractable and involves less computation than that of [17,18,19].

4.2   Characteristics of Attributes

We study the properties of attribute reducts firstly in order to obtain the necessary and sufficient conditions of attribute characteristics.

Theorem 3. *Let $T = (U, A, I)$ be a context and $D \subseteq A$ an attribute reduct. For all $a \in A$, the following statement holds:*

$$a^* \in M_U(L(T)) \Leftrightarrow \text{ there exists a unique attribute } d \in D \text{ such that } d^* = a^*.$$

Proof. Necessity. Clearly, $a^* \in M_U(L(T))$ implies $a^* \in M_U(L(T_D))$. Thus, there must exist $d \in D$ such that $d^* = a^*$. If there exists another attribute $d_1 \in D$ such that $d_1 \neq d$ and $d_1^* = a^*$, then $M_U(L(T_{D-\{d_1\}})) = M_U(L(T_D)) = M_U(L(T))$. Thus, $D - \{d_1\}$ is also an attribute consistent set of T, which is a contradiction to D is an attribute reduct. Therefore, $a^* \in M_U(L(T))$ implies there exists a unique attribute $d \in D$ such that $d^* = a^*$.

Sufficiency. We will first prove $d^* \in M_U(L(T))$ for any $d \in D$. If there exists $d \in D$ such that $d^* \notin M_U(L(T))$, then $d^* \notin M_U(L(T_D))$, which implies $M_U(L(T_D)) = M_U(L(T_{D-\{d\}}))$. Due to $M_U(L(T_D)) = M_U(L(T))$, $D - \{d\}$ is also an attribute consistent set of T, which is a contradiction. Thus, $d^* \in M_U(L(T))$ for any $d \in D$. Therefore, for any $a \in A$ if there exists an attribute $d \in D$ such that $d^* = a^*$, then $a^* \in M_U(L(T))$.

Theorem 4. *Let $T = (U, A, I)$ be a context. For all $a \in A$, the following statement holds:*

$$a^* \in M_U(L(T)) \Leftrightarrow \text{ there exists an attribute reduct } D \text{ such that } a \in D.$$

Proof. Sufficiency. The result is concluded by Theorem 3.

Necessity. If $a \notin E$ for any attribute reduct E, then there exists a unique attribute $e \in E$ such that $e^* = a^*$ by Theorem 3. Thus, $M_U(L(T_E)) = M_U(L(T_D))$, where $D = (E - \{e\}) \cup \{a\}$. Since E is an attribute reduct, it is easy to obtain that D is also an attribute reduct. The result is proved.

According to different contributions to construct the concept lattice, the attributes are classified into the following three types [17,18,19].

Definition 6. *Let (U,A, I) be a context. The set $\{D_i | D_i \subseteq A, i \in \tau\}$ (τ is an index set) includes all attribute reducts of (U, A, I). Then A is classified into the following three types:*

1. *absolute necessary attribute set $C := \bigcap_{i \in \tau} D_i$.*
2. *relative necessary attribute set $S := \bigcup_{i \in \tau} D_i - \bigcap_{i \in \tau} D_i$.*
3. *redundant attribute set $K := A - \bigcup_{i \in \tau} D_i$.*

If $a \in C$, then a is called an absolute necessary attribute; if $a \in S$, then a is called a relative necessary attribute; if $a \in K$, then a is called a redundant attribute.

By Lemma 4, Definition 5 and Definition 6, absolute necessary attributes are absolute necessary to construct the concept lattice; relative necessary attributes, together with absolute necessary attributes, determine the structure of the concept lattice; redundant attributes play no roles for the construction of the concept lattice. Now we study the characteristics of different types of attributes based on $\wedge$-irreducible elements.

Theorem 5. *Let (U,A, I) be a context. $\forall a \in A$,*

(i) *$a \in K \Leftrightarrow a^* \notin M_U(L(T))$.*
(ii) *$a \in S \Leftrightarrow a^* \in M_U(L(T))$ and there exists $a_1 \in A$, $a_1 \neq a$ such that $a_1^* = a^*$.*
(iii) *$a \in C \Leftrightarrow a^* \in M_U(L(T))$ and for any $a_1 \in A$, if $a_1 \neq a$, then $a_1^* \neq a^*$.*

Proof.

(i) According to Definition 6 and Theorem 4, we obtain

$$a \in K \Leftrightarrow a \notin D, \text{ for any attribute reduct } D,$$
$$\Leftrightarrow a^* \notin M_U(L(T)).$$

(ii) Necessity. If $a \in S$, then there exist two attribute reducts D and D_1 such that $a \in D$ and $a \notin D_1$ by Definition 6. By Theorem 3, $a^* \in M_U(L(T))$, and there must exist $a_1 \in D_1$, $a_1 \neq a$ such that $a^* = a_1^*$.
Sufficiency. The result can be concluded immediately from Theorem 3 and Theorem 4.
(iii) The result follows directly from (i) and (ii). □

Corollary. Let (U, A, I) be a context. $\forall a \in A$,

(i) a is a redundant attribute $\Leftrightarrow (a^*, a^{**})$ is not $\wedge$-irreducible.
(ii) a is a relative necessary attribute $\Leftrightarrow (a^*, a^{**})$ is $\wedge$-irreducible, and there exists $a_1 \in A$, $a_1 \neq a$ such that $a_1^* = a^*$.
(iii) a is an absolute necessary attribute $\Leftrightarrow (a^*, a^{**})$ is $\wedge$-irreducible, and for any $a_1 \in A$, $a_1 \neq a$, $a_1^* \neq a^*$.

Example 2. In Example 1, we have obtained that $M_U(L(T_{D_i})) = M_U(L(T))(i = 1, 2, 3)$, where $D_1 = \{a, b, c, e, f\}$, $D_2 = \{a, b, c, e\}$, $D_3 = \{a, c, e, f\}$. Thus, $D_i, i = 1, 2, 3$ are all attribute consistent sets of the context T in Table 1. Further, $D_2 \subsetneq D_1$ and $D_3 \subsetneq D_1$ imply D_1 is not an attribute reduct. Since $M_U(L(T_D)) \neq M_U(L(T))$ for any $D \subsetneq D_2$ or $D \subsetneq D_3$, D_2 and D_3 are attribute reducts of the context T. Combined with Lemma 4, the context T has exactly two attribute reducts D_2 and D_3. Fig. 1 and Fig. 2 show the concept lattices $L(T)$ and $L(T_{D_2})$ respectively, and it is easy to see that the two concept lattices are isomorphic to each other.

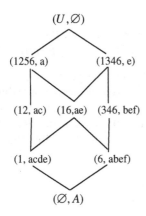

Fig. 1. The concept lattice $L(T)$

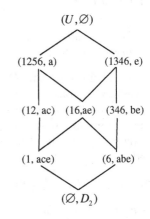

Fig. 2. The concept lattice $L(T_{D_2})$

Since $D_2 \cap D_3 = \{a, c, e\}$, $D_2 \cup D_3 = \{a, b, c, e, f\}$, we have $C = \{a, c, e\}$, $S = \{b, f\}$ and $K = \{d\}$.

4.3   Object Reduction and Characteristics in Concept Lattices

In this subsection, we study object reduction and characteristics in concept lattices based on $\vee$−irreducible elements. Since the approaches to object reduction and attribute reduction are dual, we just list the definitions and results of object reduction in concept lattices without details.

Definition 7. *Let* $T = (U, A, I)$ *be a context. An object subset* $V \subseteq U$ *is called an object consistent set of* T *if* V *satisfies*

$$J_A(L(T)) = J_A(L(T_V)). \tag{7}$$

If V is an object consistent set of T, and there is no proper subset W of V such that W is also an object consistent set of T, then V is called an object reduct of T.

This approach to object reduction is to find minimal object subsets which can preserve the intent set of $\vee$-irreducible elements in the original concept lattice.

Theorem 6. *Let $T = (U, A, I)$ be a context. For any $V \subseteq U$,*

(i) V is an object consistent set of $T \Leftrightarrow L_A(T_V) = L_A(T)$,
(ii) V is an object reduct of $T \Leftrightarrow L_A(T_V) = L_A(T)$ and $\forall x \in V$, $L_A(T_{V-\{x\}}) \neq L_A(T)$.

Theorem 6 shows that an object consistent set preserves the intent set of the original concept lattice, and then it preserves all concepts and their hierarchy of the original concept lattice. Similar with the algorithm for finding attribute reducts, we give the algorithm for obtaining object reducts based on $\vee-$irreducible elements:

Input: (U, A, I)
Output: an object reduct V
1. $Y = U$
2. $V = \emptyset$
3. **for** $(i = 1; i \leq m; i++)$ $(m = |U|)$
4. $\{x_i^* = \emptyset;$
5.   **for** $(j = 1; j \leq n; j++)$ $(n = |A|)$
6. $\{$**if** $((x_i, a_j) \in I)$ $x_i^* = x_i^* \cup \{a_j\};\}$
7. $\}$
8. **for** $(i = 1; i < m; i++)$
9. $\{$ **if** $(x_i \notin Y)$ continue;
10.   **for** $(j = i+1; j \leq m; j++)$
11. $\{$ **if** $(x_j \notin Y)$ continue;
12.   **if** $(x_i^* = x_j^*)$ $Y = Y - \{x_j\};$
13. $\}$
14. $\}$
15. Let $\{y_1, y_2, ..., y_{|Y|}\} = Y;$
16. **for** $(i = 1; i \leq |Y|; i++)$
17. $\{W_i = \emptyset;$
18.   **for** $(j = 1; j \leq |Y|; j++)$
19. $\{$**if** $(y_i^* \subseteq y_j^*$ && $y_i^* \neq y_j^*)$ $W_i = W_i \cup \{y_j\};\}$
20.   **if** $(W_i \neq \emptyset$ && $y_i^* = \bigcap_{y_j \in W_i} y_j^*)$ $V = V;$

21.   **else** $V = V \cup \{y_i\};$
22. $\}$

Then V is an object reduct of the context (U, A, I). The worst case time complexity of the algorithm is as follows. If $|Y| \leq |A|$, then the final complexity is $O(|A| \cdot |U|)$; if $|Y| > |A|$, then the final complexity is $O(|U| \cdot |Y|)$.

Theorem 7. *Let* $T = (U, A, I)$ *be a context and* V *an object reduct. For all* $x \in U$, *the following statement holds:*

$x^* \in J_A(L(T)) \Leftrightarrow$ *there exists a unique object* $y \in V$ *such that* $y^* = x^*$.

Theorem 8. *Let* $T = (U, A, I)$ *be a context. For all* $x \in U$, *the following statement holds:*

$$x^* \in J_A(L(T)) \Leftrightarrow \text{ there exists an object reduct } V \text{ such that } x \in V.$$

The objects are also classified into three types due to the different contributions to generate the concept lattice.

Definition 8. *Let* (U, A, I) *be a context. The set* $\{V_i | V_i \subseteq U, i \in \varsigma\}$ (ς *is an index set*) *includes all object reducts of* (U, A, I). *Then* U *is classified into the following three types:*

1. *absolute necessary object set* $H := \bigcap\limits_{i \in \varsigma} V_i$.
2. *relative necessary object set* $W := \bigcup\limits_{i \in \varsigma} V_i - \bigcap\limits_{i \in \varsigma} V_i$.
3. *redundant object set* $G := U - \bigcup\limits_{i \in \varsigma} V_i$.

If $x \in H$, *then* x *is called an absolute necessary object; if* $x \in W$, *then* x *is called a relative necessary object; if* $x \in H$, *then* x *is called a redundant object.*

Then the necessary and sufficient conditions of object characteristics are obtained by $\vee$-irreducible elements.

Theorem 9. *Let* (U, A, I) *be a context.* $\forall x \in U$,

(i) $x \in G \Leftrightarrow x^* \notin J_A(L(T))$.
(ii) $x \in W \Leftrightarrow x^* \in J_A(L(T))$ *and there exists* $x_1 \in U$, $x_1 \neq x$ *such that* $x_1^* = x^*$.
(iii) $x \in H \Leftrightarrow x^* \in J_A(L(T))$ *and for any* $x_1 \in U$ *if* $x_1 \neq x$, *then* $x_1^* \neq x^*$.

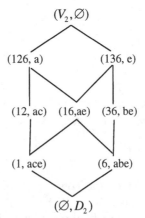

Fig. 3. The concept lattice $L(T_{D_2, V_2})$

Example 3. Combining with Example 1, we have V_1, V_2, V_3 are object consistent sets of the context T in Table 1. Because $V_j \subsetneq V_1, j = 2, 3$ and for any $V \subsetneq V_2$ or $V \subsetneq V_3$, $J_A(L(T_V)) \neq J_A(L(T))$, we know that V_1 is not an object reduct, and V_2, V_3 are both object reducts of the context T. Further, T has exactly two object reducts V_2 and V_3 according to Lemma 5 (iii), and $H = \{1, 2, 6\}, W = \{3, 4\}$ and $G = \{5\}$. Fig. 3 gives the concept lattice $L(T_{D_2, V_2})$ of the subcontext T_{D_2, V_2} after object and attribute reduction. It is obvious that the concept lattice $L(T_{D_2, V_2})$ and the original concept lattice $L(T)$ are isomorphic.

5 Conclusions

In this paper, relationships of irreducible elements between a context and its subcontext have been obtained. Based on irreducible elements, an approach to knowledge reduction including attribute reduction and object reduction in concept lattices has been developed, and the corresponding algorithms of knowledge reduction, which are more tractable and need less computation, have been provided.

Acknowledgements

The authors gratefully acknowledge the support of the National 973 Program of China (No.2002CB312200), and the Natural Science Foundation of China (No.60703117).

The authors also thank the anonymous referees for their helpful comments and suggestions.

References

1. Wille, R.: Restructuring Lattice Theory: an Approach Based on Hierarchies of Concepts. In: Rival, I. (ed.) Ordered Sets, pp. 445–470. Reidel, Dordrecht (1982)
2. Ganter, B., Wille, R.: Formal Concept Analysis: Mathematical Foundations. Springer, New York (1999)
3. Pawlak, Z.: Rough Sets. International Journal of Computer and Information Sciences 11, 341–356 (1982)
4. Kent, R.E.: Rough Concept Analysis: a Synthesis of Rough Sets and Formal Concept Analysis. Fundamenta Informaticae 27, 169–181 (1996)
5. Yao, Y.Y.: Concept Lattices in Rough Set Theory. In: 2004 Annual Meeting of the North American Fuzzy Information Processing Society (NAFIPS 2004), pp. 796–801. IEEE Press, Los Alamitos (2004)
6. Oosthuizen, G.D.: Rough Sets and Concept Lattices. In: Rough Sets, and Fuzzy Sets and Knowledge Discovery (RSKD 1993), pp. 24–31. Springer, London (1994)
7. Saquer, J., Deogun, J.: Formal Rough Concept Analysis. In: Zhong, N., Skowron, A., Ohsuga, S. (eds.) RSFDGrC 1999. LNCS(LNAI), vol. 1711, pp. 91–99. Springer, Heidelberg (1999)

8. Saquer, J., Deogun, J.: Concept Approximations Based on Rough Sets and Similarity Measures. International Journal of Applied Mathematics and Computer Sciences 11, 655–674 (2001)
9. Düntsch, I., Gediga, G.: Algebraic Aspects of Attribute Dependencies in Information Systems. Fundamenta Informaticae 29(1-2), 119–133 (1997)
10. Pagliani, P.: From Concept Lattices to Approximation Spaces: Algebraic Structures of Some Spaces of Partial Objects. Fundamenta Informaticae 18(1), 1–25 (1993)
11. Hu, K.Y., Lu, Y.C., Shi, C.Y.: Advances in Concept Lattice and Its Application. Journal of Tsinghua University (Science & Technology) 40(9), 77–81 (2000)
12. Ho, T.B.: An Approach to Concept Formation Based on Formal Concept Analysis. IEICE Transactions on Information and Systems E 782D(5), 553–559 (1995)
13. Carpineto, C., Romano, G.: GALOIS: an Order-theoretic Approach to Conceptual Clustering. In: The Tenth International Conference on Machine Learning, pp. 33–40. Elsevier, Amsterdam (1993)
14. Godin, R.: Incremental Concept Formation Algorithm Based on Galois (Concept) Lattices. Computational Intelligence 11(2), 246–267 (1995)
15. Oosthuizen, G.D.: The Application of Concept Lattice to Machine Learning. Technical Report, University of Pretoria, South Africa (1996)
16. Grigoriev, P.A., Yevtushenko, S.A.: Elements of an Agile Discovery Environment. In: Grieser, G., Tanaka, Y., Yamamoto, A. (eds.) DS 2003. LNCS(LNAI), vol. 2843, pp. 309–316. Springer, Heidelberg (2003)
17. Zhang, W.X., Wei, L., Qi, J.J.: Attribute Reduction Theory and Approach to Concept Lattice. Science in China Series F-Information Science 48(6), 713–726 (2005)
18. Zhang, W.X., Wei, L., Qi, J.J.: Attribute Reduction in Concept Lattice Based on Discernibility Matrix. In: Ślęzak, D., et al. (eds.) RSFDGrC 2005. LNCS(LNAI), vol. 3642, pp. 157–165. Springer, Heidelberg (2005)
19. Zhang, W.X., Qiu, G.F.: Uncertain Decision Making Based on Rough Sets. Tsinghua University Publishing House, Beijing (2005)
20. Carpineto, C., Romano, G.: Concept Data Analysis. John Wiley & Sons, Ltd., Chichester (2004)
21. Wang, X., Zhang, W.X.: Attribute Dependence and Reduction in Formal Contexts. International Journal of Computer Science and Knowledge Engineering 2(1), 35–49 (2008)
22. Wang, X., Ma, J.M.: A Novel Approach to Attribute Reduction in Concept Lattices. In: Wang, G., et al. (eds.) RSKT 2006. LNCS(LNAI), vol. 4062, pp. 522–529. Springer, Heidelberg (2006)
23. Wang, X., Zhang, W.X.: Relations of Attribute Reduction Between Object and Property Oriented Concept Lattices. Knowledge-Based Systems 21(5), 398–403 (2008)
24. Davey, B.A., Priestley, H.A.: Introduction to Lattices and Order. Cambridge University Press, Cambridge (2002)

A Knowledge Representation Tool
for Autonomous Machine Learning
Based on Concept Algebra

Yousheng Tian, Yingxu Wang, and Kai Hu

Theoretical and Empirical Software Engineering Research Centre (TESERC)
International Center for Cognitive Informatics (ICfCI)
Dept. of Electrical and Computer Engineering
Schulich School of Engineering, University of Calgary
2500 University Drive, NW, Calgary, Alberta, Canada T2N 1N4
Tel: (403) 220 6141; Fax: (403) 282 6855
yingxu@ucalgary.ca

Abstract. Concept algebra is an abstract mathematical structure for the formal treatment of concepts and their algebraic relations, operations, and associative rules for composing complex concepts, which provides a denotational mathematic means for knowledge system representation and manipulation. This paper presents an implementation of concept algebra by a set of simulations in Java. A visualized knowledge representation tool for concept algebra is developed, which enables machines learn concepts and knowledge autonomously. A set of eight relational operations and nine compositional operations of concept algebra are implemented in the tool to rigorously manipulate knowledge by concept networks. The knowledge representation tool is capable of presenting concepts and knowledge systems in multiple ways in order to simulate and visualize the dynamic concept networks during machine learning based on concept algebra.

Keyword: Cognitive informatics, cognitive computing, computational intelligence, denotational mathematics, intelligent systems, concept algebra, knowledge representation, machine learning, visualization, support tool, case studies, OAR, LRMB.

1 Introduction

Knowledge representation studies how to represent knowledge about the world and how to use knowledge to reason [1], [2], [7], [14], [15], [17], [18], [23], [24]. In logic, linguistics, psychology, cognitive informatics, information theory, and artificial intelligence, knowledge representation is studied as one of the important foundations for machine intelligence [4], [9], [10], [11], [12], [13], [14], [16], [21], [22]. It enables machines to abstract the real word, to reason about its environment, and to express entities and relations of the world, to model problems for efficient computation.

M.L. Gavrilova et al. (Eds.): Trans. on Comput. Sci. V, LNCS 5540, pp. 143–160, 2009.
© Springer-Verlag Berlin Heidelberg 2009

Knowledge representation is centered by the model of concepts, which is a primary element in cognitive informatics and a basic unit of thought and reasoning. Concept is an abstract model of tangible or intangible entities. Concept algebra developed by Yingxu Wang is an abstract mathematical structure for formal treatment of concepts and their algebraic relations, operations, and associative rules for composing complex concepts, which provides a denotational mathematic means for knowledge system representation and manipulation [18], [19], [20].

Concept algebra consists of three parts known as the mathematical model of abstract concepts, relational operations, and compositional operations. An abstract concept c is a prime model of concept algebra, which is defined as a 5-tuple, i.e. $c \triangleq (O, A, R^c, R^i, R^o)$ [Wang, 2008b]. The relational operations of concept algebra define relationships between concepts such as subconcept, super concept, comparison, and concept equivalence. The compositional operations of concept algebra provide a set of manipulations of abstract concepts such as inheritance, composition, and aggregation. Concept algebra provides a new mathematical structure for rigorously representing and modeling the physical and abstract worlds.

This paper presents a design and implementation of a visualization tool for knowledge representation based on concept algebra in the context of machine learning. Knowledge representation methods are analyzed in Section 2. Mappings between concepts and their relational and compositional operations in concept algebra, as well as a set of Java class simulations, are developed in Section 3. The architectural design and functional implementation of the knowledge representation tool are described in Section 4.

2   Concept Algebra for Knowledge Representation

A formal treatment of abstract concepts and a new denotational mathematical structure known as concept algebra are described in this section for knowledge representation and manipulations in cognitive informatics and computational intelligence. Before an abstract concept is introduced, the semantic environment or context [3], [5], [6], [8] in a given language, is introduced. Then, the structural model of abstract concepts, basic concept operations, and knowledge representation methods in concept algebra are formally elaborated.

A concept is an abstract structure of tangible entities or intangible ideas. As the basic unit of knowledge and reasoning, concepts can be classified into two categories known as the *concrete* and *abstract* concepts [19]. The former is a solid concept that can be directly mapped into a real-world entity or a collective set of common entities, which may be further divided into the categories of *proper* and *collective* ones. The latter is a virtual concept that cannot be directly mapped to a real-world entity. The concrete concepts are used to embody meanings of a subject in reasoning; while the abstract concepts are used as intermediate representative in reasoning.

2.1   The Abstract Model of Concepts

A *concept* is a cognitive unit to identify and/or model a real-world concrete entity and a perceived-world abstract subject.

Definition 1. Let $\mathcal{O}$ denote a finite nonempty set of *objects*, and $\mathcal{A}$ be a finite non-empty set of *attributes*, then a *semantic environment* or *context* Θ_C is denoted as a triple, i.e.:

$$\Theta_c \triangleq (\mathcal{O}, \mathcal{A}, \mathcal{R}) \tag{1}$$
$$= \mathcal{R}: \mathcal{O} \to \mathcal{O} | \mathcal{O} \to \mathcal{A} | \mathcal{A} \to \mathcal{O} | \mathcal{A} \to \mathcal{A}$$

where $\mathcal{R}$ is a set of relations between $\mathcal{O}$ and $\mathcal{A}$, and | demotes alternative relations.

Definition 2. An *abstract concept c* on Θ_C is a 5-tuple, i.e.:

$$c \triangleq (O, A, R^c, R^i, R^o) \tag{2}$$

where

- O is a finite nonempty set of objects of the concept, $O = \{o_1, o_2, ..., o_m\} \subseteq \mathbb{P}\mathcal{O}$, where $\mathbb{P}\mathcal{O}$ denotes a power set of $\mathcal{O}$.
- A is a finite nonempty set of attributes, $A = \{a_1, a_2, ..., a_n\} \subseteq \mathbb{P}\mathcal{A}$.
- $R^c = O \times A$ is a set of internal relations.
- $R^i \subseteq A' \times A$, $A' \subseteq C' \wedge A \sqsubseteq c$, is a set of input relations, where C' is a set of external concepts, $C' \subseteq \Theta_C$. For convenience, $R^i = A' \times A$ may be simply denoted as $R^i = C' \times c$.
- $R^o \subseteq c \times C'$ is a set of output relations.

Based on the above definition, an object as a concrete instantiation of a given concept can be derived as follows.

Definition 3. An *object o*, $o \in \mathbb{P}O$, as a derived bottom-level instantiation of a concept that can be mapped onto a concrete entity can be defined as follows:

$$o \triangleq (O_o, A_o, R_o^c, R_o^i, R_o^o) \tag{3}$$

where

- $O_o = \varnothing$.
- A_o is a nonempty set of attributes, $A_o = \{a_1, a_2, ..., a_n\} \subseteq \mathbb{P}\mathcal{A}$.
- $R_o^c = \varnothing$;
- $R_o^i \subseteq C' \times o$ is a set of input relations, where C' is a set of external concepts;
- $R_o^o \subseteq o \times C'$ is a set of output relations.

Definition 4. For a concept $c(O, A, R^c, R^i, R^o)$, its intension, c^*, is represented by its set of attributes A; while its extension, $c+$, is represented by its set of objects O, i.e.:

$$\begin{cases} c^* = A \\ c^+ = O \end{cases} \tag{4}$$

2.2 Concept Algebra for Knowledge Manipulations

Concept algebra is an abstract mathematical structure for the formal treatment of concepts and their algebraic relations, operations, and associative rules for composing complex concepts.

Definition 5. A *concept algebra CA* on a given semantic environment Θ_C is a triple, i.e.:

$$CA \triangleq (C, OP, \Theta_C) = ((O, A, R^c, R^i, R^o), \{\bullet_r, \bullet_c\}, \Theta_C) \tag{5}$$

where $OP = \{\bullet_r, \bullet_c\}$ are the sets of *relational* and *compositional* operations on abstract concepts.

Definition 6. The *relational operations* $\bullet_r$ in concept algebra encompass 8 comparative operators for manipulating the algebraic relations between concepts, i.e.:

$$\bullet_r \triangleq \{\leftrightarrow, \nleftrightarrow, \prec, \succ, =, \cong, \sim, \triangleq\} \tag{6}$$

where the relational operators stand for *related, independent, subconcept, superconcept, equivalent, consistent, comparison,* and *definition*, respectively.

Definition 7. The *compositional operations* $\bullet_c$ in concept algebra encompass 9 associative operators for manipulating the algebraic compositions among concepts, i.e.:

$$\bullet_c \triangleq \{\Rightarrow, \overset{-}{\Rightarrow}, \overset{+}{\Rightarrow}, \overset{\sim}{\Rightarrow}, \uplus, \Cap, \Lleftarrow, \vdash, \mapsto\} \tag{7}$$

where the compositional operators stand for *inheritance, tailoring, extension, substitute, composition, decomposition, aggregation, specification,* and *instantiation*, respectively.

Details of the relational and compositional operations of concept algebra may be referred to [19]. Concept algebra provides a denotational mathematical means for algebraic manipulations of abstract concepts. Concept algebra can be used to model, specify, and manipulate generic *"to be"* type problems, particularly system architectures, knowledge bases, and detail system designs, in cognitive informatics, computational intelligence, AI, computing, software engineering, and system engineering.

```
class concept {
        objects [] O;
        attributes [] A;
        internal_relations [] Rc;
        input_relations [] Ri;
        output_relations [] Ro;
        ...
}
```

Fig. 1. The structure of a concept class

3   Concept Manipulations by Computational Intelligence

This section describes the representations of the relational and associational manipulations of concepts in Java. The class structure of an abstract concept is modeled in Fig. 1 in Java on the basis of Eq. 2.

According to Definition 4, the methods of concept intension and extension are implemented as shown in Fig. 2, where A and O are dependent on the definition of a specific concept.

```
public attribute [] intension () {
    ...
    // A = {a₁, a₂, ..., aₙ};
    return A;
}
public concept [] extension () {
    ...
    // O = {o₁, o₂, ..., oₘ};
    return O;
}
```

Fig. 2. The intension and extension methods

3.1   Manipulations of Concept Relational Operations

In concept algebra, a set of 8 relational operations on abstract concepts, such as sub-concept, super-concept, related-concepts, independent-concepts, equivalent-concepts, consistent-concepts, concept comparison, and definition are modeled in Table 1.

Table 1. Relational Operations of Concepts in Concept Algebra

No.	Operation	Symbol	Definition	Java Implementation
1	Related	$\leftrightarrow$	$c_1 \leftrightarrow c_2 \triangleq A_1 \cap A_2 \neq \varnothing$	Fig. 3
2	Independent	$\leftrightarrow\!\!\!/$	$c_1 \leftrightarrow\!\!\!/ c_2 \triangleq A_1 \cap A_2 = \varnothing$	Fig. 3
3	Subconcept	$\prec$	$c_1 \prec c_2 \triangleq A_1 \supset A_2$	Fig. 4
4	Superconcept	$\succ$	$c_2 \succ c_1 \triangleq A_2 \subset A_1$	Fig. 4
5	Equivalent	$=$	$c_1 = c_2 \triangleq (A_1 = A_2) \wedge (O_1 = O_2)$	Fig. 5
6	Consistent	$\cong$	$c_1 \cong c_2 \triangleq (c_1 \succ c_1) \vee (c_1 \prec c_2)$ $= (A_1 \subset A_1) \vee (A_1 \supset A_2)$	Fig. 6
7	Comparison	$\sim$	$c_1 \sim c_2 \triangleq \dfrac{\#(A_1 \cap A_2)}{\#(A_1 \cup A_2)} * 100\%$	Fig. 7
8	Definition	$\triangleq$	$c_1(O_1, A_1, R^c_1, R^i_1, R^o_1) \triangleq c_2(O_2, A_2, R^c_2, R^i_2, R^o_2)$ $\triangleq c_1(O_1, A_1, R^c_1, R^i_1, R^o_1 \mid O_1 = O_2, A_1 = A_2,$ $R^c_1 = O_1 \times A_1, R^i_1 = R^i_2, R^o_1 = R^o_2)$	Fig. 8

The *related concepts* are a pair of concepts that share some common attributes in their intensions; while the independent concepts are two concepts that their intensions are disjoint. The methods for related/independent concepts are implemented in Fig. 3.

```
public boolean related_concepts (concept c1) {
        ...
    if (intersect_a (A, c1.A))
        return true;
    else
        return false;
}
  public boolean independent_concepts (concept c1) {
        ...
    if (intersect_a (A, c1.A))
            return false;
        else
            return true;
}
```

Fig. 3. Methods for related/independent concepts

The *subconcept* of a given concept is a concept that its intension is a superset; while a *superconcept* over a concept is a concept that its intension is a subset of the subconcept. The methods for sub- and super-concepts are implemented in Fig. 4.

```
public boolean sub_concept (concept c1) {
        ...
    if (inclusion_a (A, c1.A))
        return true;
    else
        return false;
}

public boolean super_concept (concept c1) {
        ...
    if (inclusion_a (c1.A, A))
        return true;
    else
        return false;
}
```

Fig. 4. Methods for sub-/super-concept operations

The *equivalent concepts* are two concepts that both of their intensions and extensions are identical. The methods for equivalent concepts are implemented in Fig. 5.

```
public boolean equivalent_concepts (concept c1) {
    …
    if (equal_a (A, c1.A)
        & equal_c (C, c1.C)
            & equal_r (Rc, c1.Rc))
            return true;
        else
            return false;
}
```

Fig. 5. The method for equivalent concepts

The *consistent concepts* are two concepts with a relation of being either a sub- or super-concept. The methods for consistent concepts are implemented in Fig. 6.

```
public boolean consistent_concepts (concept c1) {
    …
    if (related_concepts (c1)
        & sub_concept (c1)
            & super_concept (c1))
            return true;
        else
            return false;
}
```

Fig. 6. The method for consistent concepts

A *comparison* between two concepts is an operation that determines the equivalency or similarity level of their intensions. Concept comparison is implemented according to the definition in Table 1, where the range of similarity between two concepts is between 0 to 100%, in which 0% means no similarity and 100% means a full similarity between two given concepts.

```
public boolean comparison (concept c1) {
    int c_comp
    …
    C_comp = intersect_an (A, c1.A)
            / union_an (A, c1.A) *100%;
        Return c_comp;
}
```

Fig. 7. The method for comparison concepts

Concept definition is an association between two concepts where they are equivalent. The concept definition method is implemented as shown in Fig. 8.

```
public boolean concept_definition (concept c1) {
    ...
if (obj_compare(c1.O)
        & att_compare(c1.A)
        & rel_compare(Rc, c1.Rc)
        & rel_compare(Ri, c1.Ri)
        & rel_compare(Ro, c1.Ro))
        return true;
    else
        return false;
}
```

Fig. 8. The concept conjunction methods

3.2 Manipulations of Concept Compositional Operations

In concept algebra, nine concept associations have been defined as shown in Table 2 encompassing inheritance, tailoring, extension, substitute, composition, decomposition, aggregation, specification, and instantiation [19]. A set of methods are developed in this subsection to implement the compositional operations of concept algebra.

```
public boolean inheritance (concept c1) {
    ...
    if (inclusion_c (C, c1.C)
        & inclusion_a (c1.A, A)
            & inclusion_r (Rc, c1.Rc))
        return true;
        else
        return false;
}
public boolean multi_inheritance (concept[] c_n) {
    ...
    if (inclusion_c (C, c_n)
        & inclusion_a (c_n, A)
            & inclusion_r (Rc, c_n))
        return true;
        else
        return false;
}
```

Fig. 9. The inheritance and multiple inheritance methods

Table 2. Compositional Operations of Concepts in Concept Algebra

No.	Operation	Symbol	Definition	Java Implem.
1	Inheritance	$\Rightarrow$	$c_1(O_1, A_1, R^c_1, R^i_1, R^o_1) \Rightarrow c_2(O_2, A_2, R^c_2, R^i_2, R^o_2)$ $\triangleq c_2(O_2, A_2, R^c_2, R^i_2, R^o_2 \mid O_2 \subseteq O_1, A_2 \subseteq A_1, R^c_2 = O_2 \times A_2, R^i_2 = R^i_1 \cup \{(c_1, c_2)\},$ $R^o_2 = R^o_1 \cup \{(c_2, c_1)\})$ $\parallel c_1(O_1, A_1, R^c_1, R^{i'}_1, R^{o'}_1 \mid R^{i'}_1 = R^i_1 \cup \{(c_2, c_1)\}, R^{o'}_1 = R^o_1 \cup \{(c_1, c_2)\})$	Fig. 9
2	Tailoring	$\overset{\sim}{\Rightarrow}$	$c_1(O_1, A_1, R^c_1, R^i_1, R^o_1) \overset{\sim}{\Rightarrow} c_2(O_2, A_2, R^c_2, R^i_2, R^o_2)$ $\triangleq c_2(O_2, A_2, R^c_2, R^i_2, R^o_2 \mid O_2 = O_1 \setminus O', A_2 = A_1 \setminus A', R^c_2 = O_2 \times A_2 \subset R^c_1,$ $R^i_2 = R^i_1 \cup \{(c_1, c_2)\}, R^o_2 = R^o_1 \cup \{(c_2, c_1)\})$ $\parallel c_1(O_1, A_1, R^c_1, R^{i'}_1, R^{o'}_1 \mid R^{i'}_1 = R^i_1 \cup \{(c_2, c_1)\}, R^{o'}_1 = R^o_1 \cup \{(c_1, c_2)\})$	Fig. 10
3	Extension	$\overset{+}{\Rightarrow}$	$c_1(O_1, A_1, R^c_1, R^i_1, R^o_1) \overset{+}{\Rightarrow} c_2(O_2, A_2, R^c_2, R^i_2, R^o_2)$ $\triangleq c_2(O_2, A_2, R^c_2, R^i_2, R^o_2 \mid O_2 = O_1 \cup O', A_2 = A_1 \cup A', R^c_2 = O_2 \times A_2 \supset R^c_1,$ $R^i_2 = R^i_1 \cup \{(c_1, c_2)\}, R^o_2 = R^o_1 \cup \{(c_2, c_1)\})$ $\parallel c_1(O_1, A_1, R^c_1, R^{i'}_1, R^{o'}_1 \mid R^{i'}_1 = R^i_1 \cup \{(c_2, c_1)\}, R^{o'}_1 = R^o_1 \cup \{(c_1, c_2)\})$	Fig. 11
4	Substitution	$\overset{\sim}{\Rightarrow}$	$c_1(O_1, A_1, R^c_1, R^i_1, R^o_1) \overset{\sim}{\Rightarrow} c_2(O_2, A_2, R^c_2, R^i_2, R^o_2)$ $\triangleq c_2(O_2, A_2, R^c_2, R^i_2, R^o_2 \mid O_2 = (O_1 \setminus O'_{c_1}) \cup O'_{c_2}, A_2 = (A_1 \setminus A'_{c_1}) \cup A'_{c_2},$ $R^c_2 = O_2 \times A_2, R^i_2 = R^i_1 \cup \{(c_1, c_2)\}, R^o_2 = R^o_1 \cup \{(c_2, c_1)\})$ $\parallel c_1(O_1, A_1, R^c_1, R^{i'}_1, R^{o'}_1 \mid R^{i'}_1 = R^i_1 \cup \{(c_2, c_1)\}, R^{o'}_1 = R^o_1 \cup \{(c_1, c_2)\})$	Fig. 12
5	Composition	$\uplus$	$c(O, A, R^c, R^i, R^o) \uplus \overset{n}{\underset{i=1}{R}} c_i$ $\triangleq c(O, A, R^c, R^i, R^o \mid O = \bigcup_{i=1}^{n} O_{c_i}, A = \bigcup_{i=1}^{n} A_{c_i}, R^c = \bigcup_{i=1}^{n} (R^c_{c_i} \cup \{(c, c_i), (c_i, c)\},$ $R^i = \bigcup_{i=1}^{n} R^i_{c_i}, R^o = \bigcup_{i=1}^{n} R^o_{c_i})$ $\parallel \overset{n}{\underset{i=1}{R}} c_i(O_i, A_i, R^c_i, R^{i'}_i, R^{o'}_i \mid R^{i'}_i = R^i_i \cup \{(c, c_i)\}, R^{o'}_i = R^o_i \cup \{(c_i, c)\})$	Fig. 13
6	Decomposition	$\Cap$	$c(O, A, R^c, R^i, R^o) \Cap \overset{n}{\underset{i=1}{R}} c_i$ $\triangleq \overset{n}{\underset{i=1}{R}} c_i(O_i, A_i, R^c_i, R^{i'}_i, R^{o'}_i \mid R^{i'}_i = R^i_i \cup \{(c, c_i)\}, R^{o'}_i = R^o_i \cup \{(c_i, c)\})$ $\parallel c(O, A, R^c, R^i, R^o \mid O = \bigcup_{i=1}^{n} O_{c_i}, A = \bigcup_{i=1}^{n} A_{c_i}, R^c = \bigcup_{i=1}^{n} (R^c_{c_i} \setminus \{(c, c_i)(c_i, c)\}),$ $R^{i'} = R^i \cup \{\overset{n}{\underset{i=1}{R}} (c_i, c)\}, R^{o'} = R^o \cup \{\overset{n}{\underset{i=1}{R}} (c, c_i)\})$	Fig. 13
7	Aggregation/ generalization	$\Leftarrow$	$c_1(O_1, A_1, R^c_1, R^i_1, R^o_1) \Leftarrow c_2(O_2, A_2, R^c_2, R^i_2, R^o_2)$ $\triangleq c_1(O_1, A_1, R^c_1, R^i_1, R^o_1 \mid O_1 \subset O_2, A_1 \supset A_2, R^c_1 = (O_1 \times A_1) \cup \{(c_1, c_2), (c_2, c_1)\},$ $R^i_1 = R^i_2 \cup \{(c_2, c_1)\}, R^o_1 = R^o_2 \cup \{(c_1, c_2)\})$ $\parallel c_2(O_2, A_2, R^c_2, R^{i'}_2, R^{o'}_2 \mid R^{i'}_2 = R^i_2 \cup \{(c_1, c_2)\}, R^{o'}_2 = R^o_2 \cup \{(c_2, c_1)\})$	Fig. 14
8	Specification	$\vdash$	$c_1(O_1, A_1, R^c_1, R^i_1, R^o_1) \vdash c_2(O_2, A_2, R^c_2, R^i_2, R^o_2)$ $\triangleq c_2(O_2, A_2, R^c_2, R^i_2, R^o_2 \mid O_2 \supset O_1, A_2 \subset A_1, R^c_2 = (O_2 \times A_2) \cup \{(c_2, c_1), (c_1, c_2)\},$ $R^i_2 = R^i_1 \cup \{(c_1, c_2)\}, R^o_2 = R^o_1 \cup \{(c_2, c_1)\})$ $\parallel c_1(O_1, A_1, R^c_1, R^{i'}_1, R^{o'}_1 \mid R^{i'}_1 = R^i_1 \cup \{(c_2, c_1)\}, R^{o'}_1 = R^o_1 \cup \{(c_1, c)\})$	Fig. 15
9	Instantiation	$\mapsto$	$c(O, A, R^c, R^i, R^o) \mapsto o(A_o, R^c_o, R^i_o, R^o_o)$ $\triangleq o(A_o, R^c_o, R^i_o \mid o \subset O, A_o = A, R^c_o = o \times A, R^i_o = R^i_o \cup \{(c, o)\})$ $\parallel c(O, A, R^c, R^i, R^o \mid R^{i'} = R^i \cup \{(o, c)\}, R^{o'} = R^o \cup \{(c, o)\})$	Fig. 16

Concept inheritance is a concept association that indicates one concept derived from another concept. Multiple inheritances are an associative operation where a concept is derived from multiple concepts. The corresponding methods of both concept inheritance and multi-inheritance are implemented as shown in Fig. 9.

Concept tailoring as defined in Table 2 is another special inheritance operation on concepts that reduces some inherited attributes or objects in the derived concept. The implementation of the tailoring method is shown in Fig. 10.

```
public boolean tailoring (concept c1) {
    ...
    if (inclusion_c (c1.C, C)
        & inclusion_a (c1.A, A)
            & inclusion_r (c1.Rc, Rc))
        return true;
        else
        return false;
}
```

Fig. 10. The tailoring method

Concept extension as defined in Table 2 is associative operation that carries out a special concept inheritance with the introduction of additional attributes and/or objects in the derived concept. The implementation of the concept extension method is shown in Fig. 11.

```
public boolean extension (concept c1) {
    ...
    if (inclusion_c (C, c1.C)
        & inclusion_a (A, c1.A)
            & inclusion_r (Rc, c1.Rc))
        return true;
        else
        return false;
}
```

Fig. 11. The extension method

Concept substitution as defined in Table 2 is an inheritance operation that results in the replacement or overload of attributes and/or objects by locally defined ones. The implementation of the substitution method is shown in Fig. 12.

Concept composition as defined in Table 2 is associative operation that integrates multiple concepts in order to form a new complex one. The implementation of the composition method is shown in Fig. 13.

```
public boolean substitute (concept c1) {
    ...
    if (intersect_c (c1.C, C)
        & intersect_a (c1.A, A)
            & intersect_r (c1.Rc, Rc)
            & count_c (c1.C) = = count_c (C)
            & count_a (c1.A) = = count_a (A)
            & count_r (c1.Rc) = = count_r (Rc))
        return true;
    else
        return false;
}
```

Fig. 12. The substitute method

```
public boolean composition (concept[] c_n) {
    ...
    if (equal_cs (C, combine_c (c_n))
        & equal_as (A, combine_a (c_n))
            & inclusion_r (Rc, combine_r (c_n)))
        return true;
    else
        return false;
}
```

Fig. 13. The composition method

Concept aggregation as defined in Table 2 is an associative operation that assembles of complex concept by using components provided from those of multiple concepts. The implementation of the aggregation method is shown in Fig. 14.

```
public boolean aggregation (concept c1) {
    ...
    if (inclusion_c (C, c1.C)
        & inclusion_a (c1.A, A))
        return true;
    else
        return false;
}
```

Fig. 14. The aggregation method

Concept specification as defined in Table 2 is an associative operation that refines a concept by another sub-concept with more specific and precise attributes. The implementation of the specification method is shown in Fig. 15.

```
public boolean specification (concept c1) {
    ...
    if (inclusion_c (c1.C, C)
        & inclusion_a (A, c1.A))
        return true;
    else
        return false;
}
```

Fig. 15. The specification method

Concept instantiation as defined in Table 2 is a special associative operation that derives an object, or instance, on the basis of the inherited concept. The implementation of the instantiation method is shown in Fig. 16.

```
public boolean instantiation (concept c1) {
    ...
    if (inclusion_c (C, c1.C)
        & equal_as (A, c1.A)
        & (count_r (Rc) = = 0)
        & (count_r (Ro) = = 0))
        return true;
    else
        return false;
}
```

Fig. 16. The instantiation method

4   Design and Implementation of the Knowledge Representation Tool

On the basis of the mapping between concept algebra and corresponding Java components, this section describes the design and implementation of the knowledge representation system and supporting tool. Internal data structures for knowledge representation on the basis of the Object-Attribute-Relation (OAR) model [17] are introduced. Then, the architecture and functions of the tool are demonstrated.

4.1 Knowledge Representation by Concept Algebra

To implement a dynamic knowledge representation method, the OAR model is adopted. In the knowledge representation tool, a digraph model of OAR is implemented where concepts, objects, and attributes are represented by a set of nodes, and relations are represented by edges. Therefore, a knowledge system can be represented as a concept network composed of nodes and edges on the basis of the OAR model.

In a concept network, concepts can be organized in a tree or forest structure based on their abstraction levels and inheritance relationships. For example, the abstract concept *vehicle*, as shown in Fig. 17, denotes automobiles, trains, and ships, where automobiles include car and truck; trains include express train and slow train; and ships include dugout and liner. All concepts in a concept network may be categorized into different layers in the abstraction hierarchy, where the concepts in the same layer share the same abstraction level. For example, train and ship are at the same layer of automobile. However, car, truck, dugout, and liner are at a lower layer.

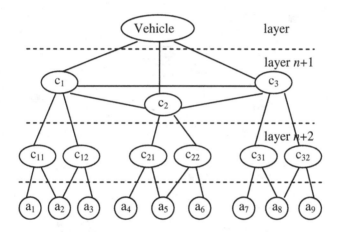

Fig. 17. An example concept network in concept algebra

Concept networks enable concepts to be efficiently accesses and manipulated. As shown in Fig. 17, $C = \{c_1, c_2, c_3\}$ = {automobile, train, ship}. Then, it is refined by: $c_1 = \{c_{11}, c_{12}\}$ = {car, truck}, $c_2 = \{c_{21}, c_{22}\}$ = {express train, slow train}, and $c_3 = \{c_{31}, c_{32}\}$ = {dugout, liner}. The set of attributes may be identified as: $A = \{a_1, a_2, a_3, a_4, a_5, a_6, a_7, a_8, a_9\}$ = {"passenger carrying", "move on road", "cargo carrier", "move on rail", "high speed", "low speed", "powered by an engine", "move in water", "use wheels"}. It is noteworthy that there may be overlapped attributes shared by multiple concepts or objects.

4.2 Design of the Knowledge Representation Tool

The knowledge representation tool is composed by the concept parser, knowledge compiler, knowledge base, and knowledge representor as illustrated in Fig. 18.

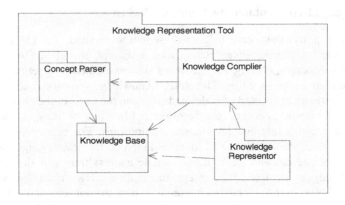

Fig. 18. The architecture of knowledge representation tool

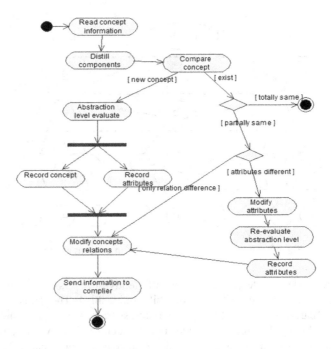

Fig. 19. An activity diagram of the concept parser

The *concept parser* identifies new concepts from the input information, which is an automatic concept gatherer. Based on given rules, it recognizes new concepts from the input information stream. When structured data streams are received, tokens such as objects, attributes, internal relations, and external relations of the concept are distilled. Identified concepts will be compared with existing concepts in the knowledge base in order to check whether it is a new concept. If it is a new concept, attributes/objects and relations will be used to decide which layer the concept belongs to.

Finally, the new concept and its attribute are stored into the concept database. Fig. 19 shows the activity diagram of the concept parser subsystem.

The *knowledge complier* recognizes new knowledge from output data streams of the concept parser. It also collects relations among new and existing concepts in order to form knowledge. Objects, attributes and layer information of given concept will be used by the knowledge complier to analyze implied relations. Fig. 20 presents the activity diagram of the knowledge complier.

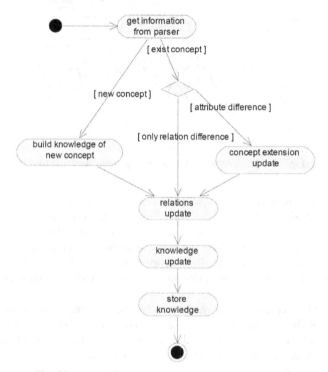

Fig. 20. An activity diagram of the knowledge complier

The *knowledge base* consists of a set of relation tables based on the OAR model [17], which incorporates the entire knowledge of acquired concepts and knowledge. The knowledge base component functions as a support center for the concept parser, the knowledge complier, and their intermediate work products.

The *knowledge representor* of the tool represents and displays knowledge in different ways such as diagrams and tables. The knowledge representor visualizes the consumed knowledge acquired by machines stored in the knowledge base.

4.3 Implementation of the Knowledge Representation Tool

Based on the system architectural as described in the preceding subsections, the knowledge representation tool has been developed in Java on Windows XP platform as shown in Fig. 21. The tool implements all concepts operations of concept algebra b a set of Java classes.

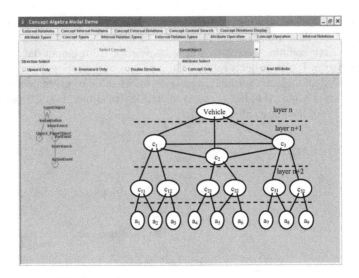

Fig. 21. The interface of the knowledge representation tool

The knowledge representation system as shown in Fig. 21 not only implements a set of functional algorithms, but also considers run-time performance, memory efficiency, and system portability. A linked digraph model is adopted to support concept operations in concept algebra. The internal knowledge structure is also flexible to guarantee query response time. Coding theories, as well as layer and inheritance information, are introduced in system design in order to effectively improve processing speed.

The knowledge representation system can be sued to transfer the concept algebra theory in denotational mathematics into a rigorous knowledge and software system design methodology. It is also adopted as a powerful tool for implementing knowledge engineering and autonomous machine learning systems.

5 Conclusions

This paper has presented the design and implementation of a knowledge representation system and supporting tool for autonomous machine learning. Concept algebra has been introduced as a novel knowledge and software system design theory and denotational mathematical means, which enables complicated knowledge and design notions be processed in a rigorous approach. Various knowledge representation methods have been analyzed, and algebraic concept operations have been implemented by Java classes. The architecture of the knowledge representation tool and its functions have been described and demonstrated. The concept algebra based methodology has been identified for a number of applications such as system requirement acquisition and elicitation in software engineering, machine learning systems, and knowledge engineering.

Acknowledgements

The authors would like to acknowledge Natural Science and Engineering Council of Canada (NSERC) for their partial support to this work. The authors would like to thank the reviewers for their valuable comments and suggestions.

References

1. Anderson, J.R.: The Architecture of Cognition. Harvard Univ. Press, Cambridge (1983)
2. Davis, R., Shrobe, H., Szolovits, P.: What is a Knowledge Representation. AI Magazine 14(1), 17–33 (1993)
3. Ganter, B., Wille, R.: Formal Concept Analysis. Springer, Berlin (1999)
4. Glasgow, J., Narayanan, N.H., Chandrasekaran, B.: Diagrammatic Reasoning: Cognitive and Computational Perspectives. AAAI Press, Menlo Park (1995)
5. Hampton, J.A.: Psychological Representation of Concepts of Memory, pp. 81–110. Psychology Press, Hove (1997)
6. Hurley, P.J.: A Concise Introduction to Logic, 6th edn. Wadsworth Pub. Co., ITP (1997)
7. Matlin, M.W.: Cognition, 4th edn. Harcourt Brace College Pub., NY (1998)
8. Medin, D.L., Shoben, E.J.: Context and Structure in Conceptual Combination. Cognitive Psychology 20, 158–190 (1988)
9. Minsky, M.: A Framework for Representing Knowledge, MIT-AI Laboratory Memo 306 (1974)
10. Sowa, J.F.: Knowledge Representation: Logical, Philosophical, and Computational Foundations. Brooks/Cole, Pacific Grove (2000)
11. Thomason, R.: Logic and artificial intelligence. The Stanford Encyclopaedia of Philosophy (2003)
12. Wang, Y.: Keynote: On Cognitive Informatics. In: Proc. 1st IEEE International Conference on Cognitive Informatics (ICCI 2002), Calgary, Canada, pp. 34–42. IEEE CS Press, Los Alamitos (2002)
13. Wang, Y.: On Cognitive Informatics. Brain and Mind: A Transdisciplinary Journal of Neuroscience and Neorophilosophy 4(3), 151–167 (2003)
14. Wang, Y.: Keynote: Cognitive Informatics - Towards the Future Generation Computers that Think and Feel. In: Proc. 5th IEEE International Conference on Cognitive Informatics (ICCI 2006), Beijing, China, pp. 3–7. IEEE CS Press, Los Alamitos (2006)
15. Wang, Y.: Software Engineering Foundations: A Software Science Perspective. CRC Book Series in Software Engineering, vol. II. Aurebach Publications, NY (2007a)
16. Wang, Y.: The Theoretical Framework of Cognitive Informatics. International Journal of Cognitive Informatics and Natural Intelligence 1(1), 1–27 (2007b)
17. Wang, Y.: The OAR Model of Neural Informatics for Internal Knowledge Representation in the Brain. International Journal of Cognitive Informatics and Natural Intelligence 1(3), 64–75 (2007c)
18. Wang, Y.: On Contemporary Denotational Mathematics for Computational Intelligence. Transactions on Computational Science 2, 6–29 (2008a)
19. Wang, Y.: On Concept Algebra: A Denotational Mathematical Structure for Knowledge and Software Modeling. International Journal of Cognitive Informatics and Natural Intelligence 2(2), 1–19 (2008b)
20. Wang, Y.: Mathematical Laws of Software. Transactions of Computational Science 2, 46–83 (2008c)

21. Wang, Y.: On Abstract Intelligence: Toward a Unified Theory of Natural, Artificial, Machinable, and Computational Intelligence. International Journal of Software Science and Computational Intelligence 1(1), 1–17 (2009a)
22. Wang, Y.: Towards a Formal Knowledge System Theory and Its Cognitive Informatics Foundations. Transactions of Computational Science 5, 1–19 (2009b)
23. Westen, D.: Psychology: Mind, Brain, and Culture, 2nd edn. John Wiley & Sons, Inc., NY (1999)
24. Wilson, R.A., Keil, F.C.: The MIT Encyclopedia of the Cognitive Sciences. MIT Press, Cambridge (2001)

Dyna: A Tool for Dynamic Knowledge Modeling

Robert Harrison and Christine W. Chan

Energy Informatics Laboratory, Faculty of Engineering
University of Regina
Regina, Saskatchewan, Canada
harrisor@uregina.ca, christine.chan@uregina.ca

Abstract. This paper presents the design and implementation of an ontology construction support tool. The Inferential Modeling Technique (IMT) (Chan, 2004), which is a technique for modeling the static and dynamic knowledge elements of a problem domain, provided the basis for the design of the tool. Existing tools lack support for modeling dynamic knowledge as defined by the IMT. Therefore, the focus of this work is development of a Protégé (Gennari, 2003) plug-in, called Dyna, which supports dynamic knowledge modeling and testing. Within Dyna, the Task Behaviour Language (TBL) supports formalized representation of the task behaviour component of dynamic knowledge. The interpreter for TBL can also enable the task behaviour representation to be run and tested, thus enabling verification and testing of the model. Dyna also supports storing the dynamic knowledge models in XML and OWL so that they can be shared and re-used across systems. The tool is applied for constructing an ontology model in the domain of petroleum contamination remediation selection.

1 Introduction

Software that enable sharing and re-use of data and functionality among distributed and heterogeneous data sources and platforms are useful because increasingly, users require functionality that can only be achieved by integrating different systems. As well, users wish to use a variety of platforms including personal computers and mobile devices. Knowledge bases typically are constructed from scratch because development of knowledge-based systems (KBS) often occurs in a distributed and heterogeneous environment involving diverse locations, system types, and knowledge representations. This makes it difficult to share and re-use existing knowledge base components. In this situation, sharing and re-use of knowledge is required in building knowledge-based systems.

The Semantic Web (SW), which is widely recognized as the next evolutionary step for the world wide web, can be used for sharing and re-use of knowledge between KBS's in a distributed and heterogeneous environment. The SW provides semantics to data on the Web, enabling computers to more easily share and perform problem solving on the data (Berners-Lee, et al., 2001). To accomplish this objective, data on the web need to be structured semantically for machine processing, and ontologies can facilitate this process of knowledge structuring. An ontology is an "explicit

M.L. Gavrilova et al. (Eds.): Trans. on Comput. Sci. V, LNCS 5540, pp. 161–181, 2009.
© Springer-Verlag Berlin Heidelberg 2009

specification of a shared conceptualization" (Gruber, 1993); it can be used for structuring the knowledge, whether it is in a KBS or on the SW. The main benefit of an ontology is that it enables the sharing and re-use of application domain knowledge across distributed and heterogeneous software systems (Guarino, 1998). Ontologies are often implemented in XML based languages, such as RDF[1] and OWL[2], which enable different KBS development groups or different Semantic Web applications to share and re-use their knowledge and data. However, construction of ontologies is difficult and time-consuming and software tools can help reduce the effort required to construct ontologies and knowledge bases.

The objective of our work is to provide software support for constructing ontologies. In this paper, we present an ontology construction support tool which specializes in providing automated support for dynamic knowledge modeling and testing; the tool is called Dyna The dual objective of providing support for modeling and testing of dynamic knowledge of an ontology is motivated by our belief that existing software tools for knowledge or ontology creation are deficient in giving support for these two areas. In order to illustrate the functionalities of Dyna, the new tool is applied for building an application ontology in the domain of petroleum contamination remediation selection.

This paper is organized as follows. Section 2 presents some background literature relevant to the field of ontology construction. Section 3 describes the development of Dyna, which is a tool for dynamic knowledge modeling and testing. Sections 4 presents the application of Dyna for developing an ontology in the petroleum contamination remediation selection domain. Section 5 presents a discussion of the tool and its application. Section 6 provides some conclusions and suggests some directions for future work.

2   Background Literature

Ontology construction support tools are software programs that enable users to formally represent an ontology while at the same time, ignore the complexities of an ontology language. They support efficient development of ontologies and help users to complete the task with fewer errors. A survey of research work on ontology construction support tools reveals a number of weaknesses common to this kind of tools (Harrison & Chan, 2005). Our research objective is to build a suite of ontology construction support tools that addresses some of these weaknesses. The following describes literature relevant to the various aspects of support necessary for ontology construction.

2.1   Ontological Engineering Methodology

The process of developing an ontology consists of activities in three different areas. First, there are ontology development activities such as specification, implementation, and maintenance. Second, there are ontology management activities such as reusing existing ontologies and quality control. And third, there are ontology support activities such as knowledge acquisition and documentation (Gomez-Perez et al., 2005). An ontological engineering methodology specifies the relationships among the activities and when the activities should be performed.

---

[1] Resource Description Framework (RDF), http://www.w3.org/RDF/
[2] OWL Web Ontology Language, http://www.w3.org/TR/owl-features/

A tool that supports an ontological engineering methodology or technique can expedite the ontological engineering process. Currently, there are few tools that directly support an ontological engineering methodology; some existing tools include: OntoEdit which supports On-To-Knowledge (Fensel et al., 2002), WebODE which supports METHONTOLOGY (Gomez-Perez et al., 2003), and the Knowledge Modeling System (KMS) which supports the Inferential Modeling Technique or IMT (Chan, 2004).

On-To-Knowledge is an iterative methodology that consists of five steps: 1. Feasibility Study, 2. Kickoff, 3. Refinement, 4. Evaluation, and 5. Maintenance (Gomez-Perez et al., 2005). OntoEdit provides support for ontology implementation, which happens in the third step of refinement, and maintenance (Fensel et al., 2002).

METHONTOLOGY is based on the software development process and provides support for the entire process during the development life cycle of an ontology. WebODE's ontology editor consists of a number of services that enable it to support various activities in the ontology life cycle defined in METHONTOLOGY. WebODE's services include ontology implementation, documentation, reasoning, and evaluation (Gomez-Perez et al., 2005).

The Inferential Modeling Technique or IMT has been developed based on the template of knowledge types advanced as the Inferential Model (Chan, 1995). The IMT supports developing a classification of the static and dynamic knowledge elements of a problem domain. Static knowledge consists of observable domain objects (classes), properties of classes, and relationships between classes. Dynamic knowledge consists of tasks (or processes) that manipulate the static knowledge to achieve an objective. The IMT is an iterative process of knowledge modeling, and the procedure is listed below. For further details on the technique, see (Chan, 2004).

1. Specify static knowledge:
   1.1 Specify the physical objects in the domain
   1.2 Specify the properties of objects
   1.3 Specify the values of the properties or define the properties as functions or equations
   1.4 Specify the relations associated with objects and properties as functions or equations
   1.5 Specify the partial order of the relations in terms of strength factors and criteria associated with the relations
   1.6 Specify the inference relations derived from the objects and properties
   1.7 Specify the partial order of the inference relations in terms of strength factors and criteria associated with the relations
2. Specify the dynamic knowledge:
   2.1 Specify the tasks
   2.2 Decompose the tasks identified into inference structures or subtasks
   2.3 Specify the partial order of the inference and subtask structures in terms of strength factors and criteria
   2.4 Specify strategic knowledge in the domain
   2.5 Specify how strategic knowledge identified is related to task and inference structures specified
3. Return to Step 1 until the specification of knowledge types is satisfactory to both the expert and knowledge engineer.

The Knowledge Modeling System (KMS) is the automated software tool developed based on the IMT for modeling static and dynamic knowledge of a problem domain (Chan 2004). KMS contains a "class" module for modeling static knowledge and a "task" module for modeling dynamic knowledge. The "class" module enables the user to model concepts or classes and properties of classes. The "class" module also supports the creation of inheritance and association relationships between classes. The "task" module enables the user to create tasks and objectives. A type of strategic knowledge can be specified by adding a prioritized list of tasks to an objective. The "task" module supports the user in linking static knowledge created in the "class" module to tasks so that task behavior or operations that manipulate objects can be defined.

The template of knowledge types in the Inferential Model and approach of IMT have been adopted in our work as the basis for our assessment of other ontology construction methodologies and tools, and Dyna was developed as an enhanced version of the KMS.

2.2 Ontology Modeling

A primary function of an ontology tool is to document the conceptual model of a problem domain. A brief survey of the different ontological engineering support tools reveal there are diverse methods employed. For example, the method for modeling an ontology used by Protégé (Gennari, 2003), Ontolingua (McGuiness et al., 2000), and OntoBuilder (Gal et al., 2006) is to use a hierarchical tree listing all the concepts relevant to a problem domain, and input fields are provided to capture characteristics of concepts. A graphical method of modeling ontologies is employed in tools such as KAON (Gabel et al., 2004) and Protégé OWLViz Plug-in (Horridge, 2005b), in which the representational method uses nodes to represent concepts and edges to represent relationships between concepts.

The tools based on the Unified Modeling Language (UML) (trademark of Object Management Group Inc., U.S.A) are commonly used to visually describe and communicate the design of software. Since ontologies are often built based on the object-oriented paradigm, UML class diagrams can be used to model ontology classes and their properties, and relationships between classes (Cranefield et al., 2003). However, UML has limited expressiveness for modeling ontologies because standard UML does not have the mechanisms for expressing descriptive logic which are needed in more advanced ontologies (Gasevic et al., 2005). To address the limitations of UML, there is on-going research work to develop a graphical notation for ontologies, called the Ontology Definition Metamodel (ODM) (Gasevic et al., 2005).

While most tools can support representation of static knowledge, they generally provide only weak support for modeling dynamic knowledge. Typically, the procedure for modeling dynamic knowledge in tools such as Protégé, KAON OI-Modeler, and KMS consists of the following three steps:

1. Model the dynamic knowledge components. In other words, classes are created to represent Task, Objective, and their properties. It should be noted that this step is not required in KMS, as it was designed for modeling dynamic knowledge and most of the necessary structures are built-in.
2. Model the static domain knowledge.

3. Use the dynamic knowledge components to create a model of the dynamic domain knowledge. In other words, the Task and Objective classes are instantiated.

We believe the existing ontology construction support tools are inadequate in providing support for modeling dynamic knowledge to the user in three aspects. First, their inability to enforce a consistent syntax for task behaviour renders computer processing of the representation of task behaviour impossible. Secondly, the lack of support for input and output of tasks can result in specification of an incomplete model. Thirdly, the visualization or input fields do not support the user in easily specifying or visualizing strategic knowledge. Dyna provides solutions for all three problems in modeling dynamic knowledge observed in existing ontology construction support tools.

2.3 Ontology Testing

The third area in the existing ontology construction support tools that requires improvement is support for ontology testing. Software testing is an important part of the software development life cycle because it helps to identify defects and can facilitate production of a more stable software application. It is also considerably cheaper to fix defects early in the development process. Unit testing is a method of testing the structure of a program. Unit testing can be used as regression tests to ensure that the software continues to work as expected after modifications. Catching and fixing defects in the modeling stage is less expensive and easier than in the implementation stage.

Ontology testing can also help the knowledge engineer develop a more complete model of the domain. In the ontological engineering field, ontology testing is also called ontology evaluation. According to (Gomez-Perez et al., 2005), ontology evaluation should be performed on the following:

- Every individual definition and axiom
- Collections of definitions and axioms stated explicitly in the ontology
- Definitions imported from other ontologies
- Definitions that can be inferred from other definitions

Existing ontology testing systems such as the OWL Unit Test Framework (Horridge, 2005a) and Chimaera's test suite (McGuiness et al., 2000) evaluate the correctness and completeness of ontologies. Chimaera performs tests in the following four areas: (1) missing argument names, documentation, constraints, etc., (2) syntactic analysis (occurrence of words, possible acronym expansion), (3) taxonomic analysis (unnecessary super classes and types), and (4) semantic evaluation (slot/type mismatch, class definition cycle, domain/range mismatch) (McGuiness et al., 2000). Such testing tools are sufficient for testing static knowledge, but are not suitable for testing the interactions between behaviour and objects, which are important in dynamic knowledge.

Our research work aims to contribute to the field of ontological evaluation by addressing the difficult issue of testing behaviour or dynamic knowledge. Our approach attempts to combine unit testing techniques with the adoption of test cases in Test

Driven Development (TDD) (Janzen & Saiedian, 2005). This is a useful hybrid approach for addressing the complex interactions of task behaviour and objects. The general intuition adopted from the TDD approach of testing is that it should be "done early and done often". In TDD, a test case is written first and then the actual module is written. The benefit of writing test cases first is that it re-orients the test designer. Instead of thinking of test cases as "how do I break something", writing test cases first makes the designer consider "what do I want the program to do". In other words, writing test cases first places the focus on defining the required functionality and objects. It is our belief that ontology development could also benefit from this approach.

Our findings from the surveyed studies provided the basis for our design of Dyna, which has been constructed so that it provides support for: (1) a knowledge modeling technique, and for this purpose, the Inferential Modeling Technique (IMT) was adopted; (2) testing of the developed knowledge model, and (3) representing the diverse types of knowledge required to model an ontology of an industrial domain in the energy and environment sector. The following discussion presents Dyna, and in order to illustrate usage of the tool, it has been applied for ontology construction in the domain of Selection of Remediation Technology for Petroleum Contamination.

3 Design and Implementation of Dyna

3.1 Overview

To address the objectives of modeling and testing dynamic knowledge, a Protégé plug-in, called Dyna, has been developed. Protégé, an ontology editing tool created by researchers at Stanford University (Gennari, 2003), is an open-source system programmed in Java and is very extensible through its plug-in architecture. Dyna is a "Tab Widget" that works with both Protégé-Frames and Protégé-OWL. At the time of writing, Dyna has been tested on Protégé 3.2.1.

A high level view of the architecture of Dyna and its interaction with Protégé is shown in Figure 1. The user creates the static knowledge of classes, properties, and relations in Protégé and the dynamic knowledge of objectives and tasks in Dyna. Dyna contains two main modules for creating objectives and tasks. From the Objective module, tasks can be linked to objectives. The Task module can support the functions of (1) definition of task behavior, (2) instantiation and manipulation of objects and properties created in Protégé, and (3) creation and running of test cases for verifying the representation of task behaviour is valid. Both Protégé and Dyna can import and export the models in the OWL file format. Dyna can also import and export the models in the XML file format.

The following describes the general process of modeling knowledge with Protégé and Dyna. A detailed description of the application process for modeling the real-world problem domain of remedial technology selection is presented in Section 4.

1. Model static knowledge in Protégé:
    1.1 Create a class
    1.2 Create a property for the class
    1.3 Specify any additional attributes for the class, such as restrictions.

    1.4 Repeat steps 1.1 – 1.3 until satisfied that the static knowledge model is complete or the knowledge engineer has enough information for progressing to Step 2 to develop the dynamic knowledge model.
2. Model dynamic knowledge in Dyna:
    2.1 Create an objective.
    2.2 Create a task
        2.2.1 Specify task behaviour
        2.2.2 Specify objects or instances of classes used in the behaviour
        2.2.3 Specify and run test cases. Object attributes can be modified and checked to see if they consist of the expected values.
        2.2.4 Link task to objective
    2.3 Repeat steps 2.1 and 2.2 until satisfied that the model is complete.
3. Optionally, the static and dynamic knowledge models can be exported in the OWL file format.

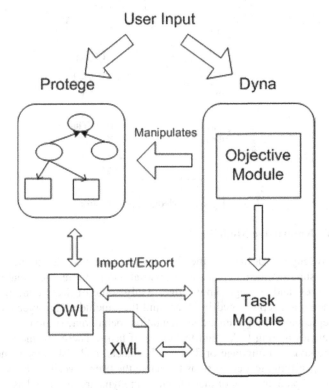

Fig. 1. Architecture of Dyna

3.2 Knowledge Representation

The static knowledge components are handled by Protégé, which uses both a Frames-based knowledge model and an OWL-based knowledge model. Both knowledge models provide classes, properties (or slots) of classes, parent-child relations between classes, and individuals (or instances) of classes. In addition, Protégé-OWL supports the many additional knowledge components defined in OWL; see http://protege.stanford.edu for more details on the Protégé knowledge models.

The dynamic knowledge components as defined by the IMT are organized into the object oriented hierarchy shown in Figure 2. The three main components of **Project**, **Task**, and **Objective** are derived from **KnowledgeComponent**. **KnowledgeComponent** provides **name** and **documentation**, which identify and describe an instance of a **KnowledgeComponent**. A **Task** is an activity that is performed to achieve an **Objective**. An **Objective** consists of a prioritized list of tasks required to achieve the objective. A **Task** consists of behaviour, input values, output values, pre-conditions, objects, and dependencies. A task can also be decomposed into sub-tasks.

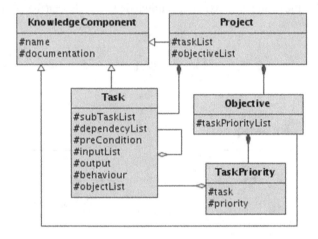

Fig. 2. Dyna knowledge representation

3.3 Dynamic Knowledge Modeling

Dynamic knowledge as defined in the IMT involves the knowledge elements of objectives, tasks, subtasks, task dependencies, partial order of inference relations, task or subtask structures, and strategic knowledge. Only some of these conceptual knowledge components are represented in Dyna, and they include: (1) objectives, (2) task knowledge represented by a name, documentation, behaviour, static objects used in a task, dependencies among tasks, and test cases of tasks. These are modeled in Dyna using Task Behaviour Language or TBL, task objects, and task dependencies. Dyna also supports specifying test cases so as to enable the user to verify the task behaviour specified. The implementation of these features in Dyna are discussed as follows.

3.3.1 Objective Modeling

Objectives are modeled in an Objective Window, which contains input fields for the name of the objective, documentation, and tasks associated with the objective. The tasks associated with an objective are in a prioritized order so that a task with higher priority should be executed first. Test cases for the tasks can also be run, and tests are run in the order of task priority. This function is discussed further in Section 3.4.

3.3.2 Task Modeling

Tasks are modeled in a Tasks Window, which contains input fields for the name of a task, documentation, behaviour, static objects used in a task, dependencies, and test cases of a task. The main components of the task module are the Task Behaviour Language (TBL) and its interpreter. The main functions of the module are (1) specification and manipulation of objects associated with a task; (2) specification and manipulation of tasks that are dependent on other tasks, and (3) support dynamic knowledge testing. The implementation mechanisms of these functions include: (a) TBL, (b) task objects, and (c) task dependency specifications; they are described as follows.

a. Task Behaviour Language

A weakness of the Knowledge Modeling System (KMS) (Chan, 2004) is that the system does not support a formal and systematic representation of task behaviour. As a result, the representation of task behaviour involves inconsistent syntax and grammar, which renders machine processing of the task behaviour representation impossible. Dyna solves this problem by using a strictly enforced, high-level language, called Task Behaviour Language (TBL), for representing task behaviour.

TBL supports the following basic structures that are common to most programming languages:

a. Types: integer, float, Boolean, string
b. Mathematical operations: +, -, *, /, %
c. Logical operators: and, or, not, xor
d. Conditional operators: <, >, ==, <=, >=, !=
a. Assignment operator: =
b. If statement, While loop, Assert statement, Return statement, Print statement
c. Class Object Instantiation
d. Function definition and calls
e. Comments: #

Dyna enforces the structure of the task behaviour with an interpreter, which also enables the task behaviour to be run. A high-level view of the architecture of the TBL Interpreter is shown in Figure 3. The interpreter consists of a lexical analyzer and a parser, which were generated using JavaCC[3], a tool for generating compilers and interpreters. The lexical analyzer breaks the input task behaviour representation into sequences of tokens. The parser then analyzes the sequences of tokens according to the TBL grammar and generates an abstract syntax tree (AST). If the input task behaviour representation has errors, then the parser outputs an error message. The functionality of each language element was implemented in the AST nodes; and there is an AST node for each TBL language structure. Interpretation of TBL is achieved by performing a depth-first traversal of the AST. As the AST is traversed, encountered symbols or identifiers are stored in the symbol table, values are pushed/popped on/off the stack, and instances of classes in Protégé are modified.

The rationale for creating TBL instead of adopting an existing language is discussed as follows. Any programming language that supports the basic structures of data types, math operators, loops, conditions, etc. could have been used for modeling the behaviour component of dynamic knowledge. The grammar that describes the C++ language for

---

[3] JavaCC, https://javacc.dev.java.net/

example, could have been processed with a tool similar to JavaCC, resulting in a lexer and parser, from which a custom interpreter could be generated. Existing programming languages have many features and language attributes, many of which are not applicable for modeling dynamic knowledge. Adopting an existing language would result in significantly more processing as the extra features need to be dealt with in some manner. Hence, it was decided that it would be easier and more efficient to create a simple language and interpreter built specifically for processing task behaviour representation. Moreover, we concluded from a survey of existing languages that non-programming languages could also have been used. For example, the Semantic Web Rule Language (SWRL)[4] is a combination of OWL and Horn-like rules, which can be used to model processes (Happel & Stojanovic, 2006). It may be possible to adopt and extend SWRL to model tasks and objectives. However, SWRL reflects an approach for modeling task knowledge that is radically different from that adopted in the IMT and KMS. Since Dyna is an enhancement of KMS, SWRL was not adopted. Instead, Dyna uses a programming-like language, called TBL, for modeling task behaviour.

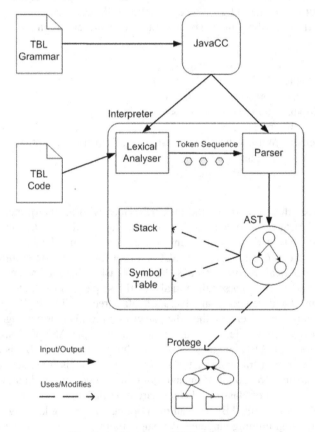

Fig. 3. Architecture of TBL Interpreter

---

b. Task Objects

Representing task behaviour in an industrial domain involves calculations and operations conducted on static knowledge. The Task Window enables the user to select a Class in Protégé to instantiate and specify a name for the resulting object. Operations and calculations can be performed on the object and its properties using TBL. For example: *car.color* = *"red"*, where *car* is an instance of a Car class, and its property *color* is set to "red". When a test case is run through the interpreter, the defined operations and calculations on the object are performed.

c. Task Dependencies

Tasks can have interactions with other tasks. When a task requires a second task to perform its behaviour, this second task is considered to be a "dependency" of the first task. Task dependencies are represented in TBL by a function called *DependentTaskName()*. Calls to dependent tasks are similar to function calls found in common programming languages, and values can be passed to the dependent task via input parameters. An output value can also be returned from the dependent task. When a dependency is encountered by the interpreter, the dependency is first added to the task's list of dependencies, and then the interpreter evaluates the behaviour of the dependency.

3.4 Testing Dynamic Knowledge

Test Driven Development (TDD) can be a very useful technique for dynamic knowledge modeling. Writing test cases for tasks first can help the ontology author better understand which classes and properties are required in the task behaviour, resulting in a more complete model. However, since writing test cases first may initially prove difficult for some users, Dyna supports the creation of test cases at any time during the ontology development process.

Dyna's testing framework is based on concepts found in JUnit[5], a unit testing framework for Java. To test a class in JUnit, a test class that contains methods representing test cases is created. A JUnit test case consists of an initialization section, a run test section, and a verification section. The initialization section defines the data required for the test case to run; the run test section is typically a call to a method that is to be tested; and the verification section contains assert functions to verify that the test is a success. Since the initialization code can be the same for multiple test cases, JUnit provides a special method called setup() for doing initialization that is common among the test cases. Similarly in Dyna, each task contains a test suite, which provides a "setup" module and facilities for creating test cases and defining them in TBL. The "setup" module is for defining initialization that is common among multiple test cases. The run test section is simply a call to the task to be tested, and the call can take input parameters and return a value. The verification section uses the "assert" function, which takes as a parameter a condition that is necessary for the test case to succeed. When the test case is run through a test interpreter, the interpreter first performs any necessary initialization, then it executes the behaviour of the task defined by the call to the task, and finally it evaluates the condition of the "assert" function. Depending on whether the "assert" function returns a "true" or "false" value, the interpreter would display a message notifying the user if the test case has "passed" or "failed". In this way, a task can be verified.

---

[5] JUnit, http://www.junit.org

4 Modeling of the Petroleum Contamination Remediation Selection Domain Using Dyna

4.1 Application Problem Domain: Petroleum Contamination Remediation

Petroleum contamination of soil and groundwater is an important environment issue because it can adversely impact the health and well-being of a community and the surrounding natural environment. Petroleum contamination is often the result of leaks and spills from petroleum storage tanks and pipelines, as shown in Figure 4. From the gas tank, contaminants first leak into the top layer of soil, then eventually through the soil, into the lower, groundwater layer. The petroleum contaminants include chemicals such as Benzene, Toluene, Ethyl Benzene, Xylene, and Total Petroleum Hydrocarbon. These chemicals can potentially cause serious health problems to humans (Chan et al., 2002).

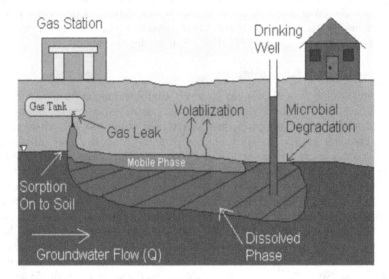

Fig. 4. Diagram of the problem of contamination from petroleum production activities (Chen, 2001)

The process of Petroleum Contamination Remediation involves a variety of remediation methods or technologies. Since different contaminated sites have different characteristics depending on the pollutants' properties, hydrological conditions, and a variety of physical (e.g. mass transfer between different phases), chemical (e.g. oxidation and reduction), and biological processes (e.g. aerobic biodegradation), the methods selected for cleaning different contaminated sites vary significantly. The decision making process for selecting a suitable method for a given contaminated site often requires expertise on both the remediation technologies and site hydrological conditions. Since the selection process is complex, an automated system for supporting decision-making on site remediation techniques is useful. Development of such a decision support system (DSS) can benefit from the use of Semantic Web technology like ontology construction (Chan et al., 2002).

4.2 Ontology Construction Process

The workflow for creating an ontology consists of a knowledge acquisition phase, followed by an ontology implementation phase. Knowledge acquisition was performed by L. L. Chen (2001), who consulted with an expert in the petroleum contamination remediation selection domain. The process of selecting a remediation technology for a petroleum contaminated site involves many steps which interact with a number of different objects. Based on the IMT, the knowledge elements in the petroleum contamination remediation selection domain are categorized into either static or dynamic knowledge. Static knowledge includes concrete objects such as soil, water, and contaminants. Dynamic knowledge includes objectives, such as "Select Remediation Method" and actions or tasks that are required to achieve the objective. Chen organized the results of the knowledge acquisition into five knowledge tables (Chen, 2001). Among these, three knowledge tables described the domain objects with their attributes, possible values, and relationships, and two knowledge tables described the dynamic knowledge. The knowledge tables were the source of information for constructing the ontologies, and both Protégé and Dyna were used in the construction process. The static knowledge was modeled in Protégé-OWL 3.2.1 and the dynamic knowledge was modeled in Dyna; the processes of ontology developed are described as follows.

4.2.1 Static Knowledge

The static knowledge of the petroleum contamination remediation selection domain was implemented in Protégé-OWL 3.2.1 using an existing ontology developed by (Chen, 2001) with the assistance of an environmental engineer (Z. Hu). Again, OWL DL was adopted because of its expressiveness and its support of automated reasoning. According to the IMT, the static knowledge was modeled using an iterative process of creating a class and then its properties. The process of class creation will be described with an example. As mentioned earlier, since the **media** (soil, water, groundwater) that has been contaminated is an important knowledge element in the Petroleum Contamination Remediation Selection domain, this element is used in the following example for illustrating the process of creating a class and other characteristics:

1. Create a class: **Media**
2. Create a property for the class. OWL supports two types of properties: datatype and object. Datatype properties are for properties of simple types such as "integer" or "string". Object properties are for properties that are individuals of other classes. The size of Media can be classified as "small", "medium", or "large". Since these are strings, a datatype property identified as **siteSize** was created.

    2.1 Specify the domain(s) of the property. For **siteSize**, the domain was set to **Media**.

    2.2 Specify the range of the property. The possible values for **siteSize** ("small", "medium", "large") are all strings, so the range of **siteSize** was set to **string**.

    2.3 Specify restrictions for the property.

    2.3.1 Restrict the possible allowed values. The only possible values for the size of a site are "small", "medium", and "large". To enforce this restriction, these values were input into the Allowed Values field.

    2.3.2 Specify whether or not the property is Functional. OWL individuals or instances can have multiple values of a property. For example a "nameList" property may have the values "John", "Jane", and "Joe". In such a situation with multiple values of a property, the property is considered to be "non-functional". If a property is limited to one value for an individual, for example a "name" property has the value "Joe", then that property is considered to be "functional". An individual of **Media** can only have one value of **siteSize** at a time (a medium cannot be both "small" and "large"), so **siteSize** is set to "functional".

3. Repeat steps 1 and 2 until either all the classes are represented or the knowledge engineer has enough information to develop the dynamic knowledge model.

4. Specify disjointness. OWL individuals can be of more than one type. For example, it is possible to specify that an individual is both a **Media** and a **Remediation**. This type of scenario may result in unforeseen negative consequences, therefore such modeling errors should be prevented. A class can be made disjoint with one or more other classes, thus avoiding the possibility of an individual being specified as belonging to more than one type. In the case of the **Media** class, it is disjoint with all of its sibling classes (**Contaminant**, **Experiment**, **Gas**, **Mathematics**, and **Remediation**).

The result of the static knowledge modeling process is the class hierarchy shown in Figure 5.

4.2.2 Dynamic Knowledge

The dynamic knowledge of the problem domain was represented in Dyna using an existing ontology by (Chen, 2001). The dynamic knowledge was modeled in Dyna and involved an iterative process of creating an objective, and then its tasks. An example of dynamic knowledge for the Petroleum Contamination Remediation Selection domain is determining the classification of the size of a site (or media). The process of creating objectives and tasks is described as follows.

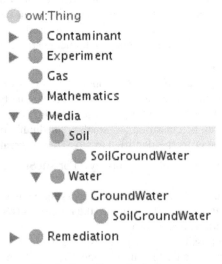

Fig. 5. Class Hierarchy in Protégé

1. Create an objective: DetermineSiteSize

Objectives are modeled in an Objective Window, which contains input fields for the name of the objective, documentation, and tasks associated with the objective. The tasks associated with an objective are in a prioritized order so that a task with higher priority should be executed first. Test cases for the tasks can also be run, and tests are run in the order of task priority.

2. Create a task: SetSiteSize

Tasks are modeled in a Tasks Window, which contains input fields for the name of a task, documentation, behaviour, objects, dependencies, and test cases of a task.

2.1   Specify task behaviour and any inputs and output. The behaviour for **Set-SiteSize** is as follows:

> *Inputs: int area, int volume*
> *if (area < 1600 and volume < 25000)*
> *{*
>   *site.siteSize = "small"*
> *}*
> *else if (area >= 1600 and area <= 2000 and*
>   *volume >= 25000 and volume <= 30000)*
> *{*
>   *site.siteSize = "medium"*
> *}*
> *else if (area > 2000 and volume > 30000)*
> *{*
>   *site.siteSize = "large"*
> *}*

2.2   This task behaviour is input into the Task Behaviour screen shown in Figure 6.

2.3   **Specify any objects, i.e. instances of classes or individuals, that are used in the behaviour**.

Task behaviour consists of calculations and operations on static knowledge represented in a class of objects. The Task Window enables the user to select a Class in Protégé, which is instantiated and for which a name is specified as the resulting object. Operations and calculations can be performed on the object and its properties using TBL. For example: *car.color = "red"*, where *car* is an instance of a Car class, and its property *color* is set to "red".

MeasureContaminatedSiteVolume uses the object **site**, which is an individual of **Media**.

2.4   Specify test cases. The testing suite, shown in Figure 7, provides facilities for adding, deleting, and running test cases. Test cases may be run individually or together. The test cases shown in Figure 7 are for the task of **Measure-ContaminatedSiteVolume**. The test case that is highlighted is for testing of a "small" site size. The test case begins with setting the area and volume of the site. Then the task, MeasureContaminatedSiteVolume, is run. Lastly, the size of the site is verified with the assert function.

3. Link the task to the objective. The link between the task **MeasureContaminated-SiteVolume** and the objective **DetermineSiteSize** is shown in Figure 8.
4. Adjust the priority of the task in the objective, if necessary. **MeasureContaminatedSiteVolume** has a priority of 2, which means that it is to be performed after the task of **MeasureContaminatedSiteArea**.
5. Repeat Steps 1-4 until the knowledge engineer is satisfied that the model is complete. While it is difficult to know when the model is complete, a useful heuristic is to assess whether all the tasks have test cases and if they run successfully. If both criteria are satisfied, then it is likely the model is complete.

```
Inputs:

Behaviour:

  # Area is in meters squared, Volume is in meters cubed

  if (site.siteArea < 1200 and site.siteVolume < 20000)
  {
      site.siteSize = "small"
      print "small"
  }
  else if (site.siteArea >= 1200 and site.siteArea <= 2000 and
              site.siteVolume >= 20000 and site.siteVolume <= 30000)
  {
      site.siteSize = "medium"
      print "medium"
  }
  else if (site.siteArea > 2000|and site.siteVolume > 30000)
  {
      site.siteSize = "large"
      print "large"
  }

Output:
```

Fig. 6. Dyna – Task Behaviour Representation

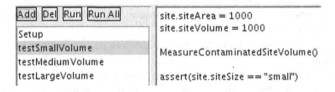

```
Add  Del  Run  Run All        site.siteArea = 1000
                              site.siteVolume = 1000
Setup
testSmallVolume               MeasureContaminatedSiteVolume()
testMediumVolume
testLargeVolume               assert(site.siteSize == "small")
```

Fig. 7. Dyna – Testing Task Behaviour

Fig. 8. Dyna – DetermineSiteSize Objective

5 Discussion

The following presents some observations about the modeling process, and about advantages and weaknesses of Dyna noted from the application modeling process.

In modeling with the IMT, a distinction is made between a "task" and an "objective". High-level operations are objectives (e.g. "SelectRemediationMethod") and operations with specific actions are tasks (e.g. "MeasureSandThickness"). Tasks can be identified using keywords that describe simple actions, such as "set", "measure", or "add"; and keywords that indicate more general or higher-level action words, such as "determine" or "find" can be used for identifying objectives. However, when this distinction is applied for modeling the remediation domain, it became apparent that the general guideline that objectives are high-level operations and tasks are specific actions cannot always be followed. For example, the knowledge table in (Chen, 2001) listed the objective "Select Remediation Method" as containing the task "Determine Site Size". Following the guideline, "Determine Site Size" is an objective as it is a high-level operation that contains a few different actions. The ideal solution would be to add the "Determine Site Size" objective to the list of tasks of "Select Remediation Method", but this action is not supported in the knowledge representation of Dyna. The solution used was to create a task called "DetermineSiteSize", of which the behaviour representation duplicates the objective's prioritized tasks. This solution is not satisfactory because the task behaviour representation duplicates the objective's prioritized tasks.

Using the correct names for tasks and objectives is very important as it greatly affects the intended meaning of the model. Many of the tasks are "Measure" tasks, such as "MeasureBenzeneConcentration". These "Measure" tasks set the value of some property for an object. In the case of "MeasureBenzeneConcentration", the value of the attribute of benzeneConcentration of the object of SoilSamplingExperiment is set. This is the typical behaviour of a "setter" function, which is used to set the value of a member variable for a class. Therefore, instead of "MeasureBenzeneConcentration", we could name the task "SetBenzeneConcentration". However, "Set" is not a satisfactory prefix in this situation because "Set" and "Measure" have different meanings. "Measure" implies that we are performing some action to find a value, while "Set" implies that we already know the value, and are simply recording it.

The IMT has suggested that a knowledge engineer should model static knowledge first, and when the static knowledge model has been developed, then the dynamic knowledge is modeled. However, we have found this suggested sequence of modeling is not always practical. Often, the ontology author is uncertain what classes and properties are required. As a consequence, creating static knowledge first leads to errors, such as incorrect names for objects are specified and properties are attached to the wrong classes. From our experience, we found a more fruitful approach is to use an iterative process of creating some dynamic knowledge first, and using it as a guide to create the static knowledge required to ensure the dynamic knowledge can be implemented. This is similar to the process of writing test cases first in Test Driven Development (TDD), described in Section 2.3. By creating tasks and objectives first, the user has to think about what functionality and objects are required. This sequence of modeling greatly reduces the amount of errors when objects are specified.

The two main components in Dyna that supported users in finding missing information and modeling errors were (1) the fields for task behaviour input and output and (2) dynamic knowledge testing. Both components were helpful in identifying new classes and properties that need to be added to the static knowledge. The fields for task behaviour input and output can be used for prompting the user to think about the relationships between tasks and the objects manipulated by the tasks. The testing of dynamic knowledge can be used for identifying new classes and properties that need to be added to the static knowledge. The creation of test cases can also reveal logical errors in the task behaviour. For example, an erroneous "if statement" for classifying site size was identified as follows:

```
if (site.siteArea < 1200 and site.siteVolume < 20000)
{
        site.siteSize = "small"
}
    else if (site.siteArea >= 1200 and site.siteArea < 2000   and site.siteVolume >=
20000 and site.siteVolume <= 30000)
{
        site.siteSize = "medium"
}
    else if (site.siteArea > 2000 and site.siteVolume > 30000)
{
        site.siteSize = "large"
}
```

The logical error is in bold and underlined in the above behaviour representation. If **site.siteArea** equals 2000, then the size of the site will not be set. Running the test cases for this task would generate a failure, thereby alerting us of the error in the model. When the error is corrected, the test cases should all run successfully. Dynamic ontology testing facilitated verification and enabled creation of a more complete knowledge model.

Not all of the dynamic knowledge in the original ontology could be modeled in Dyna. The main weakness of Dyna is its lack of support for the array data type. Many of the tasks perform operations on groups (or arrays) of objects. For example, one of the tasks measures the percentage of sand for each soil sample in the soil sampling experiment. This task requires a loop that iterates through one or more soil samples. However, since Dyna does not support arrays, only one soil sample can be measured and represented. Therefore, in implementing the ontology model, it is assumed that the number of sampling points is one (numSamplingPoints=1), and there is only one soil sample (soilSample1). This restriction can be a problem in many industrial applications of the tool.

6   Conclusions and Future Work

The objective of this work is to construct a software tool for addressing a number of weaknesses observed in existing ontology construction support systems. The weaknesses observed in existing systems such as Protégé, KAON, and KMS include their lack of support for: (1) an ontological engineering methodology, (2) dynamic knowledge modeling, and (3) dynamic knowledge testing. Dyna was designed to address these three weaknesses. To address the first problem of lack of support for an ontology engineering methodology, Dyna extended upon the functions of Protégé and included a Protégé plug-in to support the knowledge engineering method of IMT. To address the second weakness of lack of support for dynamic knowledge modeling, Dyna includes a TBL compiler for modeling task behaviour and enables enforcement of a formalized representation of task behaviour structure. The TBL representation enables task knowledge to be represented in an explicit and systematic formalism which can be easily shared and reused. To address the third weakness of lack of support for testing dynamic knowledge, Dyna supports running test cases of the task behavior specified in TBL. Dyna also supports storing dynamic knowledge representations in XML and OWL formats, so that the representations can be shared and re-used by other systems.

Dyna has been applied for constructing an ontology in the petroleum remediation selection domain. The process of modeling the petroleum contamination remediation selection domain reveals the importance of naming conventions when knowledge elements need to be labeled. Certain keywords are helpful in identifying tasks and objectives. The name of a task or objective could change the intended semantics of the knowledge element. The modeling process also reveals some advantages and weaknesses of Dyna. A strong advantage of Dyna is its support for dynamic knowledge testing, which was helpful in eliminating logical errors and in creating a more complete knowledge model. The main weakness of Dyna is the lack of support for arrays in TBL.

At time of writing, Dyna has only been applied for modeling one industrial problem domain. It is possible that with more future applications, other weaknesses of the tool will emerge. Dyna is part of a set of tools under development at the Energy

Informatics Laboratory (Harrison & Chan, 2007). Since Dyna does not currently support ontology management, a possible direction for future work is to integrate an ontology management tool with Dyna. Another direction of future work is to conduct more in-depth evaluation of Dyna. At time of writing, there are plans to conduct a preliminary evaluation of Dyna by comparing the completeness of the petroleum contamination remediation selection ontology constructed with Dyna with that built with KMS. Further research is needed to develop an appropriate formal evaluation method for assessing the ontology construction support tool.

Acknowledgements

We are grateful for the generous support of Research Grants from the Canada Research Chair Program and Natural Sciences and Engineering Research Council (NSERC). We would also like to thank Zhiying Hu for her help in clarifying knowledge in the ontology model of the domain of selection of a remediation technology for a petroleum contaminated site, and to Daniel Obst for useful discussions at various stages during the tool development process.

References

1. Berners-Lee, T., Hendler, J., Lassila, O.: The Semantic Web. Scientific American (May 2001)
2. Chan, C.W.: From Knowledge Modeling to Ontology Construction. Int. Journal of Software Engineering and Knowledge Engineering 14(6) (December 2004)
3. Chan, C.W., Huang, G., Hu, Z.: Development of an Expert Decision Support System for Selection of Remediation Technologies for Petroleum-Contaminated Sites, Final Report for Petroleum Technology Research Center, Project No. 00-03-018, University of Regina, Regina, SK, Canada (May 2002)
4. Chan, C.W.: Development and Application of A Knowledge Modeling Technique. Journal of Experimental and Theoretical Artificial Intelligence 7(2), 217–236 (1995)
5. Chen, L.L.: Construction of an ontology for the domain of selecting remediation techniques for petroleum contaminated sites, M.A.Sc. Thesis, University of Regina, Canada (2001)
6. Cranefield, S., Pan, J., Purvis, M.: A UML Ontology and Derived Content Language for a Travel Booking Scenario. In: OAS 2003, pp. 55–62 (2003)
7. Das, A., Wu, W., McGuinness, D.: Industrial Strength Ontology Management. In: Proceedings of SWWS 2001, The First Semantic Web Working Symposium, August 2001, pp. 17–38 (2001)
8. Fensel, D., Harmelen, F., van Ding, Y., Klein, M., Akkermans, H., Broekstra, J., Kampman, A., van der Meer, J., Studer, R., Sure, Y., Davies, J., Duke, A., Engels, R., Iosif, V., Kiryakov, A., Lau, T., Reimer, U., Horrocks, I.: On-To-Knowledge in a Nutshell. IEEE Computer (2002)
9. Gabel, T., Sure, Y., Voelker, J.: KAON- An Overview (April 7, 2004)
10. Gal, A., Eyal, A., Roitman, H., Jamil, H., Anaby-Tavor, A., Modica, G., Enan, M.: OntoBuilder (2006), http://iew3.technion.ac.il/OntoBuilder/
11. Gasevic, D., Djuric, D., Devedzic, V.: Ontology Modeling and MDA. Journal of Object Technology 4(1), 109–128 (2005)
12. Gennari, J.H., Musen, M.A., Fergerson, R.W., Grosso, W.E., Crubezy, M., Eriksson, H., Noy, N.F., Tu, S.W.: The evolution of Protégé: an environment for knowledge-based systems development. International Journal of Human-Computer Studies 58(1), 89–123 (2003)

13. Gomez-Perez, A., Fernandez-Lopez, M., Corcho, O.: Ontological Engineering: with examples from the areas of Knowledge Management, e-Commerce, and the Semantic Web, pp. 112, 179. Springer, Heidelberg (2005)
14. Gomez-Perez, A., Fernandez-Lopez, M., Corcho, O.: WebODE Ontology Engineering Platform (2003), http://webode.dia.fi.upm.es/WebODEWeb/index.html
15. Gruber, T.: Towards Principles for the Design of Ontologies Used for Knowledge Sharing. In: Guarino, Poli (eds.) Formal Ontology in Conceptual Analysis & Knowledge Representation. Kluwer, Padova (1993)
16. Guarino, N.: Formal Ontology in Information Systems. In: Guarino (ed.) 1st International Conference on Formal Ontology in Information Systems (FOIS 1998), Trento, Italy, pp. 3–15. IOS Press, Amsterdam (1998)
17. Harrison, R., Chan, C.W.: Tools for Industrial Knowledge Modeling. In: Proceedings of the 20th Annual Canadian Conference on Electrical and Computer Engineering (CCECE 2007), Vancouver, BC, Canada, April 22-26 (2007)
18. Harrison, R., Chan, C.W.: Implementation of an Application Ontology: A Comparison of Two Tools. In: Artificial Intelligence Applications and Innovations II: Second IFIP TC12 and WG12 Conference on Artificial Intelligence Applications and Innovations (AIAI 2005), Beijing, China, September 7-9, 2005, pp. 131–143 (2005)
19. Horridge, M.: OWL Unit Test Framework (2005a), http://www.co-ode.org/downloads/owlunittest/
20. Horridge, M.: Protégé OWLViz (2005b), http://www.co-ode.org/downloads/owlviz/co-ode-index.php
21. Huang, Z., Stuckenschmidt, H.: Reasoning with Multi-Version Ontologies: a Temporal Logic Approach. In: Gil, Y., Motta, E., Benjamins, V.R., Musen, M.A. (eds.) ISWC 2005. LNCS, vol. 3729, pp. 398–412. Springer, Heidelberg (2005)
22. Janzen, D., Saiedian, H.: Test Driven Development: Concepts, Taxonomy, and Future Direction. IEEE Computer (September 2005)
23. Klein, M., Fensel, D.: Ontology Versioning on the Semantic Web. In: Proceedings of SWWS 2001, The First Semantic Web Working Symposium, August 2001, pp. 75–92 (2001)
24. McGuiness, D., Fikes, R., Rice, J., Wilder, S.: An Environment for Merging and Testing Large Ontologies. In: Proceedings of KR 2000, pp. 485–493 (2000)
25. Menezes, A.J., van Oorschot, P.C., Vanstone, S.A.: Handbook of Applied Cryptography, p. 426 (1997)
26. Norvig, P.: AAAI: Google and the Semantic, Satanic, Romantic Web (2006), http://www.nodalpoint.org/2006/07/19/aaai_google_and_the_semantic_satanic_romantic_web
27. Noy, N.F., Chugh, A., Liu, W., Musen, M.A.: A Framework for Ontology Evolution in Collaborative Environments. In: Cruz, I., Decker, S., Allemang, D., Preist, C., Schwabe, D., Mika, P., Uschold, M., Aroyo, L.M. (eds.) ISWC 2006. LNCS, vol. 4273, pp. 544–558. Springer, Heidelberg (2006)
28. Obst, D.: Distributed Framework for Knowledge Evolution, Presentation at University of Regina Grad. Student Conf., Regina, SK, Canada (2006)
29. Richardson, M., Agrawal, R., Domingos, P.: Trust Management for the Semantic Web. In: Fensel, D., Sycara, K.P., Mylopoulos, J. (eds.) ISWC 2003. LNCS, vol. 2870, pp. 351–368. Springer, Heidelberg (2003)
30. Volz, R., Oberle, D.: KAON SERVER - A Semantic Web Management System (May 2003)
31. Yildiz, B.: Ontology Evolution and Versioning: The State of the Art, Asgaard-TR-2006-3 (October 2006)

Rough Sets and Functional Dependencies in Data: Foundations of Association Reducts

Dominik Ślęzak

Infobright Inc.
Canada/Poland
slezak@infobright.com

Abstract. We investigate the notion of an association reduct. Association reducts represent data-based functional dependencies between the sets of attributes, where it is preferred that possibly smallest sets determine possibly largest sets. We compare the notions of an association reduct to other types of reducts previously studied within the theory of rough sets. We focus particularly on modeling inexactness of dependencies, which is crucial for many real-life data applications. We also study the optimization problems and algorithms that aim at searching for the most interesting approximate association reducts in data.

Keywords: Approximate Functional Dependencies, Rough Sets, Reducts.

1 Introduction

Association rules provide a useful framework for extracting knowledge from data [1,7,14,21]. In some applications, however, the attribute-value patterns may be too specific. In the analysis of microarray data, for example, one may be interested in more general relationships between the sets of genes-attributes, rather than in detailed combinations of gene expression levels [13,18]. Also for representation purposes it may be better to work with functional dependencies between the sets of attributes, as the abstractions for the families of more specific rules based on different combinations of values. Hence, we obtain the two – the rule-based and the dependency-based – layers of data-based knowledge.

Among various approaches to modeling functional dependencies [2,4,11,15] there are information and decision reducts – irreducible subsets of attributes functionally determining all other or some distinguished attributes – introduced within the theory of rough sets [23,24,25,26]. The algorithms for efficient extraction of decision reducts from data are useful in hybrid approaches to feature selection [17,38]. Decision reducts yield the families of decision rules, along the lines of the above-mentioned two-layered interpretation of data-based knowledge. In this way, decision reducts are often applied as the rule-based classifiers [3,41].

In [30] we introduced association reducts, further studied in [31,32]. An association reduct is a pair (B, C) of disjoint attribute sets such that the values of attributes in C are (possibly approximately) determined by those in B, where B

M.L. Gavrilova et al. (Eds.): Trans. on Comput. Sci. V, LNCS 5540, pp. 182–205, 2009.

cannot be reduced and C cannot be extended without losing the required level of determinism. Association reducts correspond to the sets of association rules based on combinations of values of B (C) at their left (right) sides. Extraction of most valuable association reducts from data is, however, conducted at the level of attributes, rather than at the level of detailed attribute-value descriptors.

Certainly, association reducts should be compared not only to association rules but also to other extensions of the rough set framework for representing approximate (in)dependencies (cf. [33,34]). In particular, it is interesting to refer to the notions introduced in [36,37], e.g., to so called strong components – the pairs (B, C) of disjoint, irreducible attribute sets, which determine each other.

In this paper, we recall the basics of functional dependencies and rough set reducts (Sections 2, 3), provide discernibility-based representation for association reducts (Sections 4, 5), study approximate functional dependencies (Sections 6, 7) and discuss some new optimization problems, new heuristic search algorithms, as well as the remaining challenges related to efficient extraction of approximate association reducts from data (Sections 8, 9, 10, respectively).

When comparing to [30,31,32], the main paper's theoretical contributions are the improved Boolean representation of association reducts (Theorem 2) and the extended analysis of computational complexity of optimization principles for approximate association reducts (Theorem 4). There are also other new results formulated as propositions in Section 7 (inference about approximate functional dependencies), Section 8 (other properties of the considered optimization principles), as well as Section 9 (properties of new algorithms).

2   Functional Dependencies

Out of many approaches to data representation (cf. [1,10,11,35]), let us use information systems $\mathbb{A} = (U, A)$ with attributes $a \in A$ and objects $u \in U$ [22,23]. Each attribute yields a function $a : U \to V_a$ labeling objects with a's values from V_a. For each subset $B \subseteq A$, we define B-indiscernibility relation

$$IND(B) = \{(x, y) \in U \times U : \forall_{a \in B}\ a(x) = a(y)\} \qquad (1)$$

For every $B, C \subseteq A$, we say that B determines C, denoted by $B \Rightarrow C$, iff

$$IND(B) = IND(B \cup C) \qquad (2)$$

Such relationships, widely known as functional dependencies in databases [2,39], can be also expressed using rules. Consider the rule $\alpha \Rightarrow \beta$, where α and β are conjunctions of descriptors – pairs (a, v_a), $a \in A$, $v_a \in V_a$, also referred as literals or items [1,7]. By $\|\alpha\|$ and $\|\beta\|$ we denote the sets of objects supporting α and β in $\mathbb{A}$. The rule's support and confidence are defined as follows:[1,2]

$$supp(\alpha \wedge \beta) = \frac{\|\|\alpha\| \cap \|\beta\|\|}{|U|} \qquad conf(\alpha \Rightarrow \beta) = \frac{\|\|\alpha\| \cap \|\beta\|\|}{\|\|\alpha\|\|} \qquad (3)$$

---

[1] In this paper, we denote by $|X|$ the cardinality of a finite set X.

[2] We do not use $\mathbb{A}$ in the notation of formulas based on $\mathbb{A} = (U, A)$ because it is always clear, which system $\mathbb{A}$ we refer to. For example, instead of $IND_\mathbb{A}(B)$, $\Rightarrow_\mathbb{A}$, $\|\alpha\|_\mathbb{A}$ and $supp_\mathbb{A}(\alpha \wedge \beta)$, we write $IND(B)$, $\Rightarrow$, $\|\alpha\|$ and $supp(\alpha \wedge \beta)$.

Table 1. $\mathbb{A} = (U, A)$ with 5 binary attributes, 1 ternary attribute and 7 objects

$\mathbb{A}$	a	b	c	d	e	f
u_1	1	1	1	1	1	1
u_2	0	0	0	1	1	1
u_3	1	0	1	1	0	1
u_4	0	1	0	0	0	0
u_5	1	0	0	0	0	1
u_6	1	1	1	1	1	0
u_7	0	1	1	0	1	2

It is common knowledge that the criterion (2) can be equivalently formulated as follows: $B \Rightarrow C$ holds in $\mathbb{A}$, iff for each conjunction $\wedge_{b \in B}(b, v_b)$ that occurs in $\mathbb{A}$ there must be such a unique conjunction $\wedge_{c \in C}(c, v_c)$ that the confidence $conf$ of the association rule $\wedge_{b \in B}(b, v_b) \Rightarrow \wedge_{c \in C}(c, v_c)$ equals to 1.

Example 1. Consider $\mathbb{A} = (U, A)$, $U = \{u_1, ..., u_7\}$, $A = \{a, b, c, d, e, f\}$, illustrated in Table 1. We have attribute functions $a, b, c, d, e : U \to \{0, 1\}$, $f : U \to \{0, 1, 2\}$. One of the functional dependencies holding in $\mathbb{A}$ is $\{a, b, c\} \Rightarrow \{d, e\}$. The following corresponding rules have confidence equal to 1:

$$
\begin{array}{ll}
(a, 1) \wedge (b, 1) \wedge (c, 1) \Rightarrow (d, 1) \wedge (e, 1) & \text{supported by } u_1, u_6 \\
(a, 0) \wedge (b, 0) \wedge (c, 0) \Rightarrow (d, 1) \wedge (e, 1) & \text{supported by } u_2 \\
(a, 1) \wedge (b, 0) \wedge (c, 1) \Rightarrow (d, 1) \wedge (e, 0) & \text{supported by } u_3 \\
(a, 0) \wedge (b, 1) \wedge (c, 0) \Rightarrow (d, 0) \wedge (e, 0) & \text{supported by } u_4 \\
(a, 1) \wedge (b, 0) \wedge (c, 0) \Rightarrow (d, 0) \wedge (e, 0) & \text{supported by } u_5 \\
(a, 0) \wedge (b, 1) \wedge (c, 1) \Rightarrow (d, 0) \wedge (e, 1) & \text{supported by } u_7
\end{array}
\qquad (4)
$$

3   Rough Set Reducts

Attribute reduction is one of the steps in the knowledge discovery in databases (KDD) [17,38]. It helps unless we lose attributes that are not replaceable by the others. Generally, we should look at the sets of attributes because some attributes that seem to be insignificant as single entities may start providing important information when considered together with other attributes.

Definition 1. *[23] Let $\mathbb{A} = (U, A)$ and $B \subseteq A$ be given. We say that B is an information reduct, iff $B \Rightarrow A$ and there is no proper $B' \subsetneqq B$ such that $B' \Rightarrow A$.*

In supervised learning [15,16], one often distinguishes features called decisions. The notion of an information system can then take a form of decision system $\mathbb{A} = (U, A, D)$ with disjoint sets of conditions A and decisions D.

Definition 2. *[23] Let $\mathbb{A} = (U, A, D)$ and $B \subseteq A$ be given. We say that B is a decision reduct, iff $B \Rightarrow D$ and there is no proper $B' \subsetneqq B$ such that $B' \Rightarrow D$.*

Decision reducts correspond to the sets of decision rules in the same way as information reducts correspond to the sets of association rules (where α's are based on B and β's – on A). In Definition 2, we implicitly assume $A \Rightarrow D$ because otherwise we would not be able to expect $B \Rightarrow D$ for any $B \subseteq A$. We refer to, e.g., [28,34], where decision reducts are studied subject to a lack of $A \Rightarrow D$.

In [30,31,32], we continue the comparison between rules and reducts by letting non-trivial subsets of attributes occur at both sides of $\Rightarrow$ instead of keeping the full set A at the right side of $\Rightarrow$, which is the case of information reducts.

Definition 3. *[30] Let* $\mathbb{A} = (U, A)$ *and* $B, C \subseteq A$, $B \cap C = \emptyset$, *be given. We say that the pair* (B, C) *is an association reduct, iff* $B \Rightarrow C$ *and there is neither proper* $B' \subsetneqq B$ *nor proper* $C' \supsetneqq C$, $B \cap C' = \emptyset$, *such that* $B' \Rightarrow C$ *or* $B \Rightarrow C'$.

Example 2. For Table 1, there are information reducts: $abcf$, $abdf$, $adef$, $bcdf$, $bdef$ and cef; as well as association reducts: $(\emptyset, \emptyset)$,[3] (abc, de), $(abdf, ce)$, (abf, e), (ace, bd), (acf, d), (ade, bc), (adf, c), (aef, b), (bcd, ae), (bde, ac), (bef, a), (cdf, a) and (cef, abd). It is visible that association reducts express finer knowledge. Consider, e.g., (abc, de). Among information reducts, the closest one is $abcf$, i.e., dependency $abcf \Rightarrow de$. But for determining de we do not need f, even though f is in $\mathbb{A}$'s core, i.e., it occurs in every information reduct [23,26].

It is possible to search for association reducts and some other interesting cases of functional dependencies using relatively basic information and decision reducts previously found in data [30,37]. One can, e.g., consider a decision table $\mathbb{A} = (U, A \setminus \{a\}, \{a\})$ such that $A \setminus \{a\} \Rightarrow \{a\}$, choose a decision reduct $B \subseteq A \setminus \{a\}$ corresponding to the dependency $B \Rightarrow \{a\}$ and add more attributes to its right side unless it violates the criteria for $\Rightarrow$. Analogously, if $B \subseteq A \setminus C$ has been identified as a decision reduct in decision tables $\mathbb{A} = (U, A \setminus \{a\}, \{a\})$, $A \setminus \{a\} \Rightarrow \{a\}$, for all attributes $a \in C$, then one may deduce $B \Rightarrow C$ (although this particular technique needs to be revisited for approximate functional dependencies discussed in further sections). Nevertheless, it is far more efficient to create an algorithmic framework for deriving such functional dependencies as association reducts directly from data, with no intermediate steps related to information or decision reducts. We go back to this topic in Section 9.

4 Discernibility Matrices

Let us now discuss how to build a more formal framework for characterizing association reducts. In case of information and decision reducts, it was done by means of Boolean formulas over discernibility matrices:

Definition 4. *[26] Let* $\mathbb{A} = (U, A)$, $U = \{u_1, ..., u_N\}$, $N = |U|$, *be given. By discernibility matrix* $\mathbb{M} = [M_{ij}]$ *we mean the* $N \times N$ *matrix filled with the attribute subsets* $M_{ij} \subseteq A$ *defined as* $M_{ij} = \{a \in A : a(u_i) \neq a(u_j)\}$, *for* $i, j = 1, ..., N$.

---

[3] It means that there are no constant-valued attributes in $\mathbb{A}$. – See further sections.

Table 2. Discernibility matrix $\mathbb{M}$ for information system $\mathbb{A}$ illustrated in Table 1. $\mathbb{M}$ is symmetric and we have always $M_{ii} = \emptyset$. Hence, only the lower part is shown.

U	1	2	3	4	5	6
2	abc					
3	be	ace				
4	$acdef$	$bdef$	$abcdf$			
5	$bcde$	ade	cd	abf		
6	f	$abcf$	bef	$acde$	$bcdef$	
7	adf	$bcdf$	$abdef$	cef	$abcef$	adf

Proposition 1. *[26] Let $\mathbb{A} = (U, A)$ be given. For any $B \subseteq A$, B is an information reduct, iff, for all $i, j = 1, ..., N$, $i < j$, there is $M_{ij} \neq \emptyset \Rightarrow B \cap M_{ij} \neq \emptyset$ and there is no proper subset $B' \subsetneq B$, which would hold analogous statement.*

Now, let us consider the pair of objects (u_i, u_j) and a hypothetic association reduct (B, C). If any element of M_{ij} is going to be included into C, then we need at least one element of M_{ij} included into B to keep discernibility between u_i and u_j. Otherwise, if $C \cap M_{ij} = \emptyset$, we do not need to care about (u_i, u_j).

Proposition 2. *[31] Let $\mathbb{A} = (U, A)$ be given. For any attribute subsets $B, C \subseteq A$, the pair (B, C) forms an association reduct, iff*

$$\forall_{i<j} \, C \cap M_{ij} \neq \emptyset \Rightarrow B \cap M_{ij} \neq \emptyset \tag{5}$$

and there is neither proper subset $B' \subsetneq B$ nor proper superset $C' \supsetneq C$, for which pairs (B', C) or (B, C') would hold analogous statement.

For information reducts, one can simplify $\mathbb{M}$ by removing any M_{ij} such that there is $(k, l) \neq (i, j)$ satisfying $M_{kl} \neq \emptyset$ and $M_{kl} \subseteq M_{ij}$. Indeed, if we require a hypothetic information reduct to intersect both with M_{ij} and M_{kl}, given $M_{kl} \subseteq M_{ij}$, we can equivalently require non-empty intersection only with M_{kl}. For instance, the only remaining elements after simplification of the above $\mathbb{M}$ are $M_{12}, M_{13}, M_{16}, M_{23}, M_{25}$ and M_{35}. Most of simplification power lays in the core attributes [23,26]. For above $\mathbb{M}$, $M_{16} = \{f\}$ eliminates all other M_{ij} containing f. For association reducts (B, C), such core attributes may or may not occur in B but they cannot occur in C, as illustrated by Example 2.

Discernibility matrices can be also reconsidered by means of binary tables with rows corresponding to the pairs of objects $(u_i, u_j) \in U \times U$, columns corresponding to the original attributes $a \in A$ and values notifying whether each given two objects have the same or different values on a given column. In the rough set literature, the value of 1 would usually notify us about the case of $a(u_i) \neq a(u_j)$. However, we suggest exchanging the meanings of 1 and 0, which is actually equivalent to the methodology for handling functional dependencies within multi-valued contexts in the formal concept analysis:

Table 3. Indiscernibility table $\mathbb{I}$ for Table 1. Rows (u_i, u_j) and (u_j, u_i) are identical. Rows (u_i, u_i) contain only 1's. Hence, only rows (u_i, u_j) such that $i < j$ are shown.

$U \times U$	a	b	c	d	e	f
(u_1, u_2)	0	0	0	1	1	1
(u_1, u_3)	1	0	1	1	0	1
(u_1, u_4)	0	1	0	0	0	0
(u_1, u_5)	1	0	0	0	0	1
(u_1, u_6)	1	1	1	1	1	0
(u_1, u_7)	0	1	1	0	1	0
(u_2, u_3)	0	1	0	1	0	1
(u_2, u_4)	1	0	1	0	0	0
(u_2, u_5)	0	1	1	0	0	1
(u_2, u_6)	0	0	0	1	1	0

	a	b	c	d	e	f
(u_2, u_7)	1	0	0	0	1	0
(u_3, u_4)	0	0	0	0	1	0
(u_3, u_5)	1	1	0	0	1	1
(u_3, u_6)	1	0	1	1	0	0
(u_3, u_7)	0	0	1	0	0	0
(u_4, u_5)	0	0	1	1	1	0
(u_4, u_6)	0	1	0	0	0	1
(u_4, u_7)	1	1	0	1	0	0
(u_5, u_6)	1	0	0	0	0	0
(u_5, u_7)	0	0	0	1	0	0
(u_6, u_7)	0	1	1	0	1	0

Definition 5. *(cf. [11]) Let $\mathbb{A} = (U, A)$, $U = \{u_1, ..., u_N\}$, be given. By indiscernibility table $\mathbb{I} = (U \times U, A)$ we mean the binary information system, where $a((u_i, u_j)) = 1$ iff $a(u_i) = a(u_j)$, for any $a \in A$ and $i, j = 1, ..., N$.*

Association rules with confidence equal to 1 in $\mathbb{I}$ correspond to association reducts in the original information system $\mathbb{A}$.[4] Let us also note that calculation of support and accuracy of association rules within $\mathbb{I}$ is related to quite interesting interpretations of approximate functional dependencies in $\mathbb{A}$. Certainly, similar characteristics may be considered for decision reducts [20,28]. In this study, we omit decision reducts for better clarity. In Section 5, we focus on the next step of building characteristics for association reducts, which is related to Boolean reasoning. In Section 6, we go back to approximate functional dependencies.

5   Boolean Representation

Consider Boolean function τ [5]. Product term t (conjunction of non-contradictory literals – variables or their negations) is called an implicant of τ, iff τ is true for all valuations of variables that make t true. Further, t is called a prime implicant for τ, iff it is τ's implicant and there is no proper t's subterm (with some literals removed), which would be still τ's implicant.

Theorem 1. *[26] Let $\mathbb{A} = (U, A)$ be given. Consider the following Boolean function, where every Boolean variable a is identified with attribute $a \in A$:*

$$\tau \equiv \wedge_{i,j:M_{ij} \neq \emptyset}(\vee_{a \in M_{ij}} a) \tag{6}$$

Then, every given subset $B \subseteq A$ is an information reduct for $\mathbb{A}$, iff the product term $t(B) \equiv \wedge_{a \in B} a$ is a prime implicant for τ.

---

[4] In this particular part of the paper, we refer to the original way of understanding association rules for the binary data. Elsewhere, we let the rules base on descriptors $a = v_a$, where $v_a \in V_a$ does not need to be 1 and V_a does not need to be $\{0, 1\}$.

Using absorption laws, we remove from τ the clauses corresponding to reducible sets M_{ij}. For $\mathbb{M}$ illustrated in Section 4, we get the following function with the prime implicants corresponding to reducts listed in Example 2:

$$\tau \equiv (a \vee b \vee c) \wedge (b \vee e) \wedge (f) \wedge (a \vee c \vee e) \wedge (a \vee d \vee e) \wedge (c \vee d) \tag{7}$$

In case of association reducts (B,C), we need to consider both types of requirements (5) within a Boolean function: non-empty intersections of B with M_{ij} and, otherwise, empty intersections of C with M_{ij}. Therefore, we use the two following types of Boolean variables, corresponding to every $a \in A$:

1. Variable a is true, iff attribute a belongs to B
2. Variable a^* is true, iff attribute $a \in A$ does not belong to C

It leads to the following result pre-formulated in [31], refined here in CNF form:

Theorem 2. *Let* $\mathbb{A} = (U, A)$ *be given. Consider the following Boolean function:*

$$\tau^* \equiv \wedge_{i,j:M_{ij} \neq \emptyset}(\wedge_{X:X \subseteq M_{ij}, X \neq \emptyset}((\vee_{a \in M_{ij} \setminus X} a) \vee (\vee_{a \in X} a^*))) \tag{8}$$

Then, for every $B, C \subseteq A$, $B \cap C = \emptyset$, (B,C) *forms the association reduct, iff there is the following prime implicant* $t(B,C)$ *for* τ^*:

$$t(B,C) \equiv (\wedge_{a \in B} a) \wedge (\wedge_{a \notin C} a^*) \tag{9}$$

Proof. We sketch the proof and provide further illustration in Appendix A.

Example 3. These are the prime implicants of τ^* for $\mathbb{A}$ illustrated in Table 1: $a^*b^*c^*d^*e^*f^*$, $abca^*b^*c^*f^*$, $abdfa^*b^*d^*f^*$, $abfa^*b^*c^*d^*f^*$, $acea^*c^*e^*f^*$, $acfa^*b^* c^*e^*f^*$, $adea^*d^*e^*f^*$, $adfa^*b^*d^*e^*f^*$, $aefa^*c^*d^*e^*f^*$, $bcdb^*c^*d^*f^*$, $bdeb^*d^*e^*f^*$, $befb^*c^*d^*e^*f^*$, $cdfb^*c^*d^*e^*f^*$ and $cefc^*e^*f^*$. One can verify that they correspond to the association reducts in Example 2. For instance, $acea^*c^*e^*f^*$ corresponds to the pair (B,C), where $B = \{a,c,e\}$ and $C = A \setminus \{a,c,e,f\} = \{b,d\}$.

Please note that the length of each of the obtained prime implicants is directly proportional to the cardinality of C and inversely proportional to the cardinality of B. It fits with a general idea of searching for most interesting reducts as those corresponding to the shortest prime implicants (cf. [20,26]).

6   Approximate Dependencies

While extracting association rules $\alpha \Rightarrow \beta$ from data, we focus on those with shortest α and longest β. The same should be the case for functional dependencies $B \Rightarrow C$, as further discussed while formulating the optimization problems in Section 8. Another issue is introducing more flexibility with regards to the dependencies' precision, since in real-world data we can count on fully exact relationships neither at the level of the sets of attributes (in case of reducts) nor combinations of their specific values (in case of rules).

Generally, one can imagine a set of approximation thresholds Θ corresponding to different levels of θ-approximate dependencies in a given $\mathbb{A} = (U, A)$:

$$B \Rightarrow_\theta C \qquad B, C \subseteq A, \ \theta \in \Theta \tag{10}$$

For every $\theta \in \Theta$, it should be assumed that θ-approximate dependencies satisfy some laws. In our research, we are especially interested in the following ones:

$$\begin{aligned} \text{IF} \quad X \Rightarrow_\theta Y \cup Z \quad \text{THEN} \quad X \Rightarrow_\theta Y \\ \text{IF} \quad X \Rightarrow_\theta Y \quad \text{THEN} \quad X \cup Z \Rightarrow_\theta Y \end{aligned} \tag{11}$$

Properties (11) imply that $X \Rightarrow_\theta \emptyset$, as well as IF $X \Rightarrow_\theta Y \cup Z$ THEN $X \cup Z \Rightarrow_\theta Y$. It may also happen that for some $X \subseteq A$ there is $\emptyset \Rightarrow_\theta X$, denoted shortly by $_\theta X$. Then we may say that attributes in X are θ-constant or θ-regular.

In applications, the analysis of multiple approximation thresholds is useful. We assume an ordering $\preceq$ over Θ, with the following property satisfied:

$$\text{IF} \quad \theta \preceq \vartheta \quad \text{THEN} \quad X \Rightarrow_\theta Y \Rightarrow X \Rightarrow_\vartheta Y \tag{12}$$

Generally, although all further examples refer to natural and real numbers, we assume that Θ is a complete lattice [6]. In particular, we assume the following:

$$\forall_{\Theta' \subseteq \Theta} \exists_{\theta' \in \Theta} [\forall_{\vartheta \in \Theta'} \theta' \preceq \vartheta \wedge \forall_{\theta \in \Theta} (\forall_{\vartheta \in \Theta'} \theta \preceq \vartheta \Rightarrow \theta \preceq \theta')] \tag{13}$$

Such $\theta' \in \Theta$ is called the meet of Θ', denoted as $\bigwedge \Theta'$. In particular, we assume that the least Θ's element $\mathbf{0} = \bigwedge \Theta$ corresponds to exact dependencies (2): [5]

$$X \Rightarrow_\mathbf{0} Y \Leftrightarrow X \Rightarrow Y \tag{14}$$

We consider triplets $(\Theta, \preceq, \Rightarrow_\theta)$ as the means for analyzing families of approximate dependencies (10). Properties (11,12) form a framework for inference about the approximate dependencies. Certainly, such a framework differs from the inference rules originally proposed for functional dependencies in [2]. Properties (11,12) and (14) refer also to the way of generalizing association reducts onto the approximate case. Before proceeding with such generalization in Section 7, we provide below some examples of approximate dependencies.

Let us start by coming back to association rules, where approximations are regulated by the confidence thresholds. One of the most popular tasks is to search for association rules with $conf(\alpha \Rightarrow \beta) \geq t$, under additional constraint $supp(\alpha \wedge \beta) \geq s$, for some $t, s \in (0, 1]$. In case of functional dependencies, one could think about the thresholds reflecting average confidence and support of the corresponding rules. One of the functions reflecting such averages is the information entropy [10], defined for each $B \subseteq A$ as follows:[6]

$$H(B) = \sum_{\wedge_{b \in B}(b, v_b): \ supp(\wedge_{b \in B}(b, v_b)) > 0} supp(\wedge_{b \in B}(b, v_b)) \log(supp(\wedge_{b \in B}(b, v_b))) \tag{15}$$

---

[5] We focus by default on the nominal attributes. However, the meaning of $X \Rightarrow Y$, $X \Rightarrow_\mathbf{0} Y$ and equivalence (14) can be reconsidered for other data types as well.

[6] In this paper, all logarithms have the base equal to 2.

$H(B)$ corresponds to the support of patterns generated by B over U [29]:

$$H(B) = -\log\left(\prod_{u \in U} supp(\wedge_{b \in B}(b, b(u)))\right)^{\frac{1}{|U|}} \tag{16}$$

Similarly, conditional entropy $H(C|B) = H(B \cup C) - H(B)$ equals to

$$H(C|B) = -\log\left(\prod_{u \in U} conf(\wedge_{b \in B}(b, b(u)) \Rightarrow \wedge_{c \in C}(c, c(u)))\right)^{\frac{1}{|U|}} \tag{17}$$

Definition 6. *(cf. [27,30]) Consider $\Theta = [0, +\infty)$, $\mathbf{0} = 0$. Let us put:*

$$X \Rightarrow_\theta^H Y \;\Leftrightarrow\; H(Y|X) \leq \theta \tag{18}$$

It is common knowledge that $\Rightarrow_\theta^H$ satisfies (11,12,14) [10]. H has been already thoroughly studied within the rough set framework (cf. [9,29]), given the first attempts to use it for modeling approximate dependencies in [27].

Example 4. In Example 1, we considered $abc \Rightarrow de$, which can be now rephrased as $abc \Rightarrow_\theta^H de$ for $\theta = 0$. We can see that $H(abc) = H(abcde) = \log 7 - 2/7$, so $H(de|abc) = 0$. We can also see by means of entropy that $abc \Rightarrow_\theta^H de$ cannot be strengthened (by removing attributes from $\{a, b, c\}$ or adding f to $\{d, e\}$) without increasing the value of θ. For instance, we have $H(def|abc) = 2/7$.

Definition 7. *(cf. [21,28]) Consider $\Theta = \mathbb{N}$, $\mathbf{0} = 0$. Let us put:*

$$B \Rightarrow_\theta^D C \;\Leftrightarrow\; DISC(C|B) \leq \theta \tag{19}$$

where $DISC(C|B) = DISC(B \cup C) - DISC(B)$ and

$$DISC(B) = |\{(x, y) \in U \times U : \exists_{a \in B}\, a(x) \neq a(y)\}| \tag{20}$$

Again, properties (11,12,14) are satisfied. $\Rightarrow_\theta^D B$ means that the number of pairs in $\{(x, y) \in U \times U : \exists_{a \in B}\, a(x) \neq a(y)\}$ does not exceed θ. $DISC(B)$ corresponds to arithmetic average of supports of patterns generated on B, similarly to the case of entropy related to their geometric average. Criteria analogous to (19) have been already studied by means of discernibility, decision rules' supports, satisfaction of "almost all" clauses in Boolean formulas etc. (cf. [19,28]).

Example 5. Consider dependency $abc \Rightarrow_\theta^D de$ for $\theta = 0$. We have $DISC(abc) = DISC(abcde) = 40$, so $DISC(de|abc) = 0$. As in Example 6, there is inequality $DISC(def|abc) > 0$. Actually, the approximate dependency $abc \Rightarrow_\theta^D def$ would hold for $\theta = 2$ or higher. As we observe similar behavior of $DISC$ and H in this case, let us look closer at possible ranges of H and $DISC$. For data in Table 1, H can vary from 0 to $\log 7$ (the value of $H(A)$), while $DISC$ – from 0 to 42 (the value of $DISC(A)$). Hence, while tuning parameter θ within available ranges, somebody's judgement of the degree of dependency of $\{d, e, f\}$ from $\{a, b, c\}$ may turn out to be quite different for H ($2/7 \in [0, \log 7]$) and $DISC$ ($2 \in [0, 42]$).

Definition 8. *Consider $\Theta = [0, 1)$, $\mathbf{0} = 0$. Let us put:*

$$B \Rightarrow_\theta^I C \;\Leftrightarrow\; INDS(C|B) \geq 1 - \theta \tag{21}$$

where $INDS(C|B) = INDS(B \cup C)/INDS(B)$ and

$$INDS(B) = |\{(x, y) \in U \times U : \forall_{a \in B}\, a(x) = a(y)\}| \tag{22}$$

The quantities of $INDS(B \cup C)/|U|^2$ and $INDS(C|B)$ are equal, respectively, to the support and confidence of association rules corresponding to the pairs (B, C) in $\mathbb{I}$ introduced in Definition 5. There is also a relationship between the quantities of $DISC$ and $INDS$ leading, e.g., to equality $DISC(C|B) = INDS(B) - INDS(B \cup C)$. On the other hand, although the approximate dependency criteria $\Rightarrow_\theta^D$ and $\Rightarrow_\theta^I$ have so similar foundations, $\Rightarrow_\theta^I$ does not satisfy (11). It is a kind of difficulty because of importance of (11) while generalizing association reducts onto the case of approximate dependencies in Section 7.

As a summary, there are many models of approximate functional dependencies, which may lead to different characteristics. In further sections, in examples, we will restrict mostly to H. However, all the following considerations and results remain valid also for other settings of $(\Theta, \preceq, \Rightarrow_\theta)$ satisfying properties (11-14).

7   Approximate Reducts

The notions of rough set reducts can be now reformulated in terms of approximate dependencies. Such an approach has its origin in [27], though it received more recognition after publishing [24], where it was written as follows: *The idea of an approximate reduct can be useful in cases when a smaller number of condition attributes is preferred over accuracy of classification.* This statement is worth rephrasing for a broader variety of applications, not only those related to classification – a domain where decision reducts/rules play a crucial role within the rough set framework. Further, it is not always about using a smaller number of attributes, as – in case of association reducts – we are also interested in maximizing the number of attributes at the right sides of dependencies. Generally, the following principle may be considered: *We may agree to slightly decrease the model's consistency with data, if it leads to significantly more powerful knowledge representation.* Such principle seems to be valid not only for reducts but also for their ensembles treated as complex classifiers or the means for inference about functional dependencies (cf. [33,41]), as well as for other models of approximate relationships among attributes (cf. [13,34]). In this study, we restrict to the analysis of approximate information/association reducts.

Definition 9. *Consider $(\Theta, \preceq, \Rightarrow_\theta)$, $\mathbb{A} = (U, A)$, $\theta \in \Theta$ and $B \subseteq A$. We say that B is a θ-information reduct, iff $B \Rightarrow_\theta A$ and there is no proper $B' \subsetneq B$ such that $B' \Rightarrow_\theta A$. Further, let $B, C \subseteq A$, $B \cap C = \emptyset$, be given. We say that (B, C) is a θ-association reduct, iff we have $B \Rightarrow_\theta C$ and there is neither proper $B' \subsetneq B$ nor proper $C' \supsetneq C$, $B \cap C' = \emptyset$, such that $B' \Rightarrow_\theta C$ or $B \Rightarrow_\theta C'$.*

Properties (11) enable us to generalize the well-known characteristics of decision/information reducts [23,26] in Proposition 3. It is important to realize that these properties are necessary for each given framework $(\Theta, \preceq, \Rightarrow_\theta)$ to have such characteristics. For further examples of the approximate reduction criteria (defined for decision reducts but usually with a straightforward way of introducing their information/association analogies), we refer, e.g., to [28,42].

Proposition 3. *Consider $(\Theta, \preceq, \Rightarrow_\theta)$ and $\mathbb{A} = (U, A)$. Assume properties (11). $B \subseteq A$ is a θ-information reduct, iff $B \Rightarrow_\theta A$ and there is no $b \in B$ such that $B \setminus \{b\} \Rightarrow_\theta A$. Similarly, (B, C) is a θ-association reduct, iff $B \Rightarrow_\theta C$ and there are neither $b \in B$ such that $B \setminus \{b\} \Rightarrow_\theta C$ nor $c \notin B \cup C$ such that $B \Rightarrow_\theta C \cup \{c\}$.*

Proof. Straightforward application of (11).

Definition 10. *[30] Let $\theta \geq 0$, $\mathbb{A} = (U, A)$ and $B, C \subseteq A$, $B \cap C = \emptyset$, be given. We say that (B, C) is an (H, θ)-association reduct, iff $B \Rightarrow_\theta^H C$ and there is neither $B' \subsetneq B$ nor $C' \supsetneq C$, $B \cap C' = \emptyset$, such that $B' \Rightarrow_\theta^H C$ or $B \Rightarrow_\theta^H C'$.*

Example 6. Consider $\mathbb{A} = (U, A)$ in Table 1, for $\theta = 1/2$, as well as $B = \{e, f\}$ and $C = \{a, b\}$. Given $H(ab|ef) = 2/7$, we have $ef \Rightarrow_{1/2}^H ab$. In order to verify whether (ef, ab) is an (H, θ)-association reduct for $\theta = 1/2$, we need to check $H(ab|e)$, $H(ab|f)$, $H(abc|ef)$ and $H(abd|ef)$. All those conditional entropies are greater than $1/2$. Thanks to the properties of H, it means that $H(ab|\emptyset)$ ($H(ab)$ in other words) and $H(abcd|ef)$ are greater than $1/2$ too.

Definition 9 can be reformulated without an explicit requirement of setting up particular $\theta \in \Theta$. It is important because the choice of approximation level is usually not so obvious for the data sets related to specific applications.

Definition 11. *Consider $(\Theta, \preceq, \Rightarrow_\theta)$ and $\mathbb{A} = (U, A)$. We say that the pair $(B, \theta) \in 2^A \times \Theta$ is a $*$-information reduct, iff B is a θ-information reduct and there is no $\theta' \npreceq \theta$ such that B is a θ'-information reduct. Further, we say that $(B, C, \theta) \in 2^A \times 2^A \times \Theta$ is a $*$-association reduct, iff (B, C) is a θ-association reduct and there is no $\theta' \npreceq \theta$ such that (B, C) is a θ'-association reduct.*

Intuitively, $*$-association reducts are of the highest importance while analyzing approximate functional dependencies regardless of agreeable levels of approximation. We discuss it further in the next sections. Let us conclude this part by going back to the idea of interpreting (11) and (12) as the inference rules.

Proposition 4. *Consider $(\Theta, \preceq, \Rightarrow_\theta)$ and $\mathbb{A} = (U, A)$. Assume (11,12). Consider $\theta \in \Theta$ and $B, C \subseteq A$ such that $B \Rightarrow_\theta C$ holds in $\mathbb{A}$. $B \Rightarrow_\theta C$ is derivable from the set of θ-association reducts by the inference rules (11). It is also derivable from the set of $*$-association reducts by the inference rules (11,12).*

Proof. One can easily verify that for each case of $B \Rightarrow_\theta C$ there exists at least one θ-association reduct (B', C') such that $B' \subseteq B$ and $C' \supseteq C$. Similarly, there exists at least one $*$-association reduct (B', C', θ') such that, additionally, $\theta' \preceq \theta$. Hence, the proof of Proposition 4 is a straightforward application of (11,12).

8   Optimization Problems

According to Proposition 4, the set of all association reducts can be treated as the basis for inference about approximate functional dependencies in data (cf. [32,33]). However, one cannot operate with such large sets in practice. The number of decision and information reducts is proven to be exponentially large with respect to the number of attributes (cf. [20,26]). The same is expected for association reducts, although a deeper analysis with this respect is still one of our future goals (see also Appendix A). In our research, we rather focus on searching for reducts yielding maximum knowledge. Analogous strategy might be applied also to other approaches to representation of and inference about functional dependencies (cf. [2,4]). In our opinion, replacing complete basis with significantly smaller ensembles of dependencies providing reasonably rich knowledge about data is one of the most important tasks in this area. Below, we discuss several optimization problems referring to extraction of the most valuable association reducts and, next, the most valuable subsets of association reducts from data.

Definition 12. *[32] F-Optimal Θ-Dependency Problem (FΘ) for $(\Theta, \preceq, \Rightarrow_\theta)$ and function $F : \mathbb{N} \times \mathbb{N} \to \mathbb{R}$. Input: $\theta \in \Theta$; $\mathbb{A} = (U, A)$. Output: If there are any pairs $B, C \subseteq A$, such that $B \Rightarrow_\theta C$, $B \cap C = \emptyset$ and $(B, C) \neq (\emptyset, \emptyset)$, then return (B, C) with maximum value of $F(|B|, |C|)$; Otherwise, return $(\emptyset, \emptyset)$.*

Function F should reflect the objective of searching for dependencies $B \Rightarrow_\theta C$ with the smallest B and largest C. It is reasonable to assume the following:

$$n < n' \Rightarrow F(n, m) > F(n', m) \quad \wedge \quad m < m' \Rightarrow F(n, m) < F(n, m') \quad (23)$$

Proposition 5. *[32] Consider $(\Theta, \preceq, \Rightarrow_\theta)$ and $F : \mathbb{N} \times \mathbb{N} \to \mathbb{R}$. Assume (11) and (23). For any $\theta \in \Theta$ and $\mathbb{A} = (U, A)$, FΘ results in a θ-association reduct.*

The choice of F, e.g. $F(n, m) = m - n$ or $F(n, m) = m/(n + 1)$, may certainly influence the resulting association reducts. On the other hand, one would like to investigate the properties of FΘ for a possibly large family of functions. We analyze the complexity of FΘ for an arbitrary F which satisfies (23).

Theorem 3. *[32] If $(\Theta, \preceq, \Rightarrow_\theta)$ holds (14) and F holds (23), FΘ is NP-hard.*

Considering cardinalities as the arguments of F is only one of possibilities, just like in case of decision and information reducts, where optimization may involve the number of generated rules, their average supports, etc. (cf. [29,41]). Such an idea can be also addressed by rephrasing the criteria for association rules constrained by their supports [1,7]. For example, we may search for (H, θ)-association reducts (B, C) such that $H(B \cup C) \leq \vartheta$ holds for some $\vartheta \geq \theta \geq 0$.

Definition 13. *Consider $(\Theta, \preceq, \Rightarrow_\theta)$ and $\mathbb{A} = (U, A)$. Let $\theta, \vartheta \in \Theta$, $\theta \preceq \vartheta$, and $B, C \subseteq A$ be given. (B, C) is called a (θ, ϑ)-dependency, iff*

$$B \Rightarrow_\theta C \quad \wedge \quad {}_\vartheta(B \cup C) \quad (24)$$

(B, C) is called a (θ, ϑ)-association reduct, iff it satisfies (24), there is $B \cap C = \emptyset$, and there are neither (θ, ϑ)-dependencies (B', C), where $B' \subsetneq B$, nor (θ, ϑ)-dependencies (B, C'), where $C' \supsetneq C$ and $B \cap C' = \emptyset$.

Example 7. Unless ϑ is set up high, the pairs (abc, de) and (ef, ab) considered in Examples 1 and 5 should be disregarded. This is because $H(abcde) = H(abef) = \log 7 - 2/7$, i.e., the supports of the corresponding rules are very poor.

Definition 14. *F-Optimal Θ_2-Dependency Problem ($F\Theta_2$) for $(\Theta, \preceq, \Rightarrow_\theta)$ and function $F : \mathbb{N} \times \mathbb{N} \rightarrow \mathbb{R}$. Input: $\theta, \vartheta \in \Theta$, $\theta \preceq \vartheta$; $\mathbb{A} = (U, A)$. Output: If there are any pairs $B, C \subseteq A$, which satisfy (24), $B \cap C = \emptyset$ and $(B, C) \neq (\emptyset, \emptyset)$, then return (B, C) with maximum value of $F(|B|, |C|)$; Otherwise, return $(\emptyset, \emptyset)$.*

Proposition 6. *Consider $(\Theta, \preceq, \Rightarrow_\theta)$ and $F : \mathbb{N} \times \mathbb{N} \rightarrow \mathbb{R}$. Assume (23). For any $\theta, \vartheta \in \Theta$ and $\mathbb{A} = (U, A)$, $F\Theta_2$ results in a (θ, ϑ)-association reduct.*

Proof. Assume that the resulting (B, C) is not a (θ, ϑ)-association reduct. It would mean that there is a pair (B', C') satisfying (24) and such that $B' \subsetneq B$ or $C' \supsetneq C$. According to (23), it would mean that $F(|B|, |C|) < F(|B'|, |C'|)$.

Theorem 4. *If $(\Theta, \preceq, \Rightarrow_\theta)$ holds (14) and F holds (23), $F\Theta_2$ is NP-hard.*

Proof. We sketch the proof in Appendix B.

Analogous analysis of the problems introduced in Definitions 16 and 17 is one of our future goals. Let us first refer to Definition 11 and to the following notion:

Definition 15. *Consider $(\Theta, \preceq, \Rightarrow_\theta)$ and $\mathbb{A} = (U, A)$. We say that $(B, C, \theta, \vartheta) \in 2^A \times 2^A \times \Theta \times \Theta$ is a $(*, *)$-association reduct, iff (B, C) is a (θ, ϑ)-association reduct and there is no $(\theta', \vartheta') \preceq (\theta, \vartheta)$, where $\theta' \not\succeq \theta$ or $\vartheta' \not\succeq \vartheta$, such that (B, C) is a (θ', ϑ')-association reduct.*

The idea behind $*$-association reducts and $(*, *)$-association reducts is to look at the most valuable pairs (B, C) without referring to any pre-defined constraints for approximation thresholds. If we consider a task of searching for dependencies $B \Rightarrow_\theta C$ that are particularly useful for inference about the others, we should look not only at $|B|$ and $|C|$ but also at a potential of using the inference rule (12) for thresholds $\theta' \in \Theta$ such that $\theta \preceq \theta'$. In case of $(*, *)$-association reducts, one can interpret the meaning of $\vartheta \in \Theta$ similarly. It leads to the following assumptions for optimization function $F^* : \mathbb{N} \times \mathbb{N} \times \Theta \times \Theta \rightarrow \mathbb{R}$:

$$
\begin{aligned}
n < n' &\Rightarrow F^*(n, m, \theta, \vartheta) > F^*(n', m, \theta, \vartheta) \\
m < m' &\Rightarrow F^*(n, m, \theta, \vartheta) < F^*(n, m', \theta, \vartheta) \\
\theta \not\succeq \theta' &\Rightarrow F^*(n, m, \theta, \vartheta) > F^*(n, m, \theta', \vartheta) \\
\vartheta \not\succeq \vartheta' &\Rightarrow F^*(n, m, \theta, \vartheta) > F^*(n, m, \theta, \vartheta')
\end{aligned}
\tag{25}
$$

Definition 16. *F^*-Optimal Θ_2^*-Dependency Problem ($F^*\Theta_2^*$) for $(\Theta, \preceq, \Rightarrow_\theta)$ and $F^* : \mathbb{N} \times \mathbb{N} \times \Theta \times \Theta \rightarrow \mathbb{R}$. Input: $\mathbb{A} = (U, A)$. Output: $(B, C, \theta, \vartheta) \in 2^A \times 2^A \times \Theta \times \Theta$ with maximum value of $F^*(|B|, |C|, \theta, \vartheta)$.*

Proposition 7. *Consider $(\Theta, \preceq, \Rightarrow_\theta)$ and $F^* : \mathbb{N} \times \mathbb{N} \times \Theta \times \Theta \rightarrow \mathbb{R}$. Assume (25). For any $\mathbb{A} = (U, A)$, $F^*\Theta_2^*$ results in a $(*, *)$-association reduct.*

Proof. Analogous to the proof of Proposition 6.

The above ideas may be surely reconsidered for $*$-information and $*$-association reducts. In the same way, the discussion below does not need to restrict only to (θ, ϑ)-association reducts. It is crucial to note that Definition 17 illustrates two independent aspects of approximation related to functional dependencies: dealing with approximate functional dependencies $\Rightarrow_\theta$ and, on top of that, tending towards possibly compact and complete representation of approximate knowledge about (approximate or exact) functional dependencies.

Definition 17. *Optimal k-Set Θ_2-Dependency Problem $(\Theta_2\#_k)$ for $k \in \mathbb{N}$ and $(\Theta, \preceq, \Rightarrow_\theta)$ satisfying (11). Input: $\theta, \vartheta \in \Theta$, $\theta \preceq \vartheta$; $\mathbb{A} = (U, A)$. Output: At most k (θ, ϑ)-dependencies, which enable to derive maximum amount of other (θ, ϑ)-dependencies and which cannot be derived from each other by (11).*

Proposition 8. *For $k \in \mathbb{N}$ and $(\Theta, \preceq, \Rightarrow_\theta)$ satisfying (11), for $\theta, \vartheta \in \Theta$, $\theta \preceq \vartheta$, and $\mathbb{A} = (U, A)$, $\Theta_2\#_k$'s output consists of $\min\{k, R_{\theta, \vartheta}\}$ (θ, ϑ)-association reducts, wherein $R_{\theta, \vartheta}$ denotes the number of (θ, ϑ)-association reducts in $\mathbb{A}$.*

Proof. Consider (θ, ϑ)-dependency (B, C) that is not a (θ, ϑ)-association reduct. There is at least one (θ, ϑ)-association reduct (B', C') such that $B' \subseteq B, C' \supseteq C$, and $B' \subsetneq B$ or $C' \supsetneq C$. All (θ, ϑ)-dependencies derivable from (B, C) are derivable from (B', C'). Moreover, in the particular case of inference rules (11), (B', C') can be derived neither from (B, C) nor from other (θ, ϑ)-dependencies. Hence, regardless of other pairs of attribute subsets in $\Theta_2\#_k$'s output, (B', C') will always contribute with more additionally derivable (θ, ϑ)-dependencies than (B, C) would do. Let us also note that the set of all (θ, ϑ)-association reducts enables to derive all (θ, ϑ)-dependencies. Hence, if $R_{\theta, \vartheta} < k$, then $\Theta_2\#_k$'s output will consist of all (θ, ϑ)-association reducts. Any additional (θ, ϑ)-dependencies would be then redundant, violating the requirement for elements of $\Theta_2\#_k$'s output to be non-derivable from each other. Otherwise, if $R_{\theta, \vartheta} \geq k$, then $\Theta_2\#_k$'s output will include as many (θ, ϑ)-association reducts as possible, that is k.

Definition 17 requires comparison with the previous methodologies based on the sets of reducts. However, the main attention so far was paid on the classification models based on the sets of decision reducts [33,41] – an example of a popular research stream related to the classifier ensembles [15,16]. $\Theta_2\#_k$ refers to the inference rather than to the classification abilities of approximate association reducts. Let us note that in Definition 17 there is no reference to the cardinalities of the reducts' components. Actually, as emphasized also in the next section, optimal sets of association reducts (B, C) may contain not necessarily the reducts with maximum values of $F(|B|, |C|)$ but rather the reducts complementing each other in providing maximum knowledge about dependencies.

The problems analogous to $\Theta_2\#_k$ can be formulated for other schemes of inference about functional dependencies (cf. [2,4]) as well as for other types of functional dependency models (cf. [13,34]), not necessarily related to the specific case of association reducts and inference rules limited to (11). In particular, when applying the set of inference rules originally proposed in [2], it may happen that some of association reducts can be inferred from the others.

9   Algorithmic Framework

There are numerous approaches to searching for reducts in data [3,42]. In particular, in case of association reducts, one can adapt some algorithms designed to search for association rules [1,7] to work at the level of approximate functional dependencies. For example, in case of $\Rightarrow_\theta^I$ introduced in Definition 8, one may search for association rules in the indiscernibility table $\mathbb{I}$, with their support and confidence complying with the notion of a (θ, ϑ)-association reduct in Definition 13. We can further modify those algorithms to work directly with information system $\mathbb{A}$ and compute the values of $INDS$ with no need of constructing $\mathbb{I}$. We can extend the original algorithms onto the cases of other approximate dependencies, e.g. $\Rightarrow_\theta^D$ or $\Rightarrow_\theta^H$. Finally, as proposed for $\Rightarrow_\theta^H$ in [30], we can modify the algorithms to take an advantage of the second out of the implications in (11), which has no analogy at the level of association rules.

In this paper, we focus on methodology which is not adopted from the realm of association rules but it rather extends one of the approaches developed for decision reducts [40]. The basic idea is to work with permutations of attributes, generated randomly or conducted by the order-based genetic algorithm (o-GA, [8]). Permutation (individual in o-GA) σ launches a fast deterministic algorithm resulting in a reduct of a specific type, further evaluated with respect to a considered optimization criterion. In case of o-GA, the result of evaluation is assigned to σ as its fitness. We discuss several of such permutation-based procedures for different types of reducts introduced in the previous sections. Let us start with a relatively simple case of θ-information reducts.

Proposition 9. *[32] Let $\mathbb{A} = (U, A)$ and $\theta \in \Theta$ satisfying (11) be given. 1) For every permutation, the output of Algorithm 1 is a θ-information reduct. 2) For every θ-information reduct, there exists a permutation giving it as the output.*

The above result means that all solutions can be reached and that all outcomes of Algorithm 1 satisfy the constraints of irreducibility of a reduct. It enables the whole search process to better focus on the optimization criteria. In o-GA, the fitness of individual σ may be defined, e.g., as $1/(|B|+1)$ or as $F(|B|, |A \setminus B|)$, for F considered in Definition 12. Regardless of whether we generate permutations purely randomly or use mechanisms of o-GA, the overall outcome of the search process is the best-found θ-information reduct or reducts. The most valuable reducts are reachable by a higher percentage of permutations [33,41]. Therefore, the process is likely to quickly converge towards nearly optimal solutions.

Example 8. Consider $\theta = \mathbf{0}$ (14). Recall information reducts in Example 2. Note that cef is the shortest reduct and, additionally, it focuses on attributes slightly less frequent in other reducts. Such a "uniqueness" of cef makes it reachable by more permutations. Out of $6! = 720$ permutations, 264 lead to cef; 96 to each of $abcf$, $abdf$, $adef$; and 84 to each of $bcdf$, $bdef$. Attributes a, c, e occur in the existing reducts three times; b, d four times; and f six times. Out of a, c, e only a does not belong to the "strongest" reduct cef. It makes reducts $abcf$, $abdf$ and $adef$ reachable by more permutations than in case of $bcdf$ and $bdef$.

Algorithm 1. θ-information reduct calculation for $(\Theta, \preceq, \Rightarrow_\theta)$ and $\mathbb{A} = (U, A)$

Input: $\theta \in \Theta$; $\sigma : \{1, ..., n\} \to \{1, ..., n\}$, $n = |A|$
Output: $B \subseteq A$

  $B \leftarrow A$
  for $i = 1$ to n **do**
    if $B \setminus \{a_{\sigma(i)}\} \Rightarrow_\theta A$ **then**
      $B \leftarrow B \setminus \{a_{\sigma(i)}\}$
    end if
  end for
  return B

In [32], we provided a similar procedure for θ-association reducts. Here, we re-formulate it as Algorithm 2 for the (θ, ϑ)-association case. There are two phases:

1. We follow σ, assembling possibly large $B \subseteq A$ satisfying $_\vartheta B$ (one may claim here a partial analogy to the first phase of Apriori algorithm [1]; cf. [30])
2. We follow σ again, now removing as much as possible from B and adding as much as possible to C, subject to the constraints $B \Rightarrow_\theta C$ and $_\vartheta(B \cup C)$

Proposition 10. *Let $\mathbb{A} = (U, A)$ and $(\Theta, \preceq, \Rightarrow_\theta)$ satisfying (11,12) be given. For arbitrary $\theta, \vartheta \in \Theta$, $\theta \preceq \vartheta$ and permutation $\sigma : \{1, ..., n\} \to \{1, ..., n\}$, the result of Algorithm 2 is a (θ, ϑ)-association reduct.*

Proof. $B \cap C = \emptyset$ and (24) are obvious. Further, according to (11), we cannot remove anything from the left side of the final result because if we cannot remove a given $a_{\sigma(i)}$ from B in (*), we would not be able to do it any time later, after more removals from B and additions to C. Similarly, we cannot add anything to the right side of the final result, unless we could do it earlier in (**).

Unfortunately, it is not said that each (θ, ϑ)-association reduct is reachable by some σ. Still, we have a kind of convergence towards valuable solutions:

Example 9. Let us go back to association reducts as defined in Section 3, i.e., consider $\theta = 0$ and $\vartheta \in \Theta$ such that $_\vartheta A$. Out of 14 association reducts listed in Example 2, only seven occur as the outputs of Algorithm 2. 96 permutations lead to (abc, de), 60 to $(abdf, ce)$, 84 to (ace, bd), 132 to (ade, bc), 96 to (bcd, ae), 114 to (bde, ac) and 138 to (cef, abd). We may say that association reducts with similar left/right sides "compete" with each other for the same areas of the space of permutations. This is the case of, e.g., (ace, bd) and (ade, bc). However, a significant amount of permutations that might lead to (ace, bd) is additionally "stolen" by (cef, abd). Such behavior of permutations is consistent with the fact how the sets of functional dependencies derivable from (ace, bd) and (cef, abd) overlap. On the other hand, association reducts based on more unique attribute sets are reachable by more permutations and, in the same time, yield functional dependencies that are less frequently derivable from other reducts. It makes Algorithm 2 potentially useful also in resolving optimization problems related to the sets of association reducts, like e.g. $\Theta_2 \#_k$ introduced in Definition 17.

Algorithm 2. (θ, ϑ)-association reduct calculation for $(\Theta, \preceq, \Rightarrow_\theta)$ and $\mathbb{A} = (U, A)$

Input: $\theta, \vartheta \in \Theta$, $\theta \preceq \vartheta$; $\sigma : \{1, ..., n\} \to \{1, ..., n\}$
Output: $(B, C) \in 2^A \times 2^A$

    $B \leftarrow \emptyset$, $C \leftarrow \emptyset$
    for $i = 1$ to n **do**
        if $_\vartheta(B \cup \{a_{\sigma(i)}\})$ **then**
            $B \leftarrow B \cup \{a_{\sigma(i)}\}$
        end if
    end for   // the first phase ends here
    for $i = 1$ to n **do**
        if $B \setminus \{a_{\sigma(i)}\} \Rightarrow_\theta C$ **then**
            $B \leftarrow B \setminus \{a_{\sigma(i)}\}$ // (*) here we decrease B
            if $B \Rightarrow_\theta C \cup \{a_{\sigma(i)}\}$ **then**
                if $_\vartheta(B \cup C \cup \{a_{\sigma(i)}\})$ **then**
                    $C \leftarrow C \cup \{a_{\sigma(i)}\}$ // (**) here we increase C
                end if
            end if
        end if
    end for   // the second phase ends here
    return (B, C)

There are two aspects of complexity of the presented heuristic search methodology based on permutations: complexity of the procedures such as Algorithm 2 performed for each single permutation, as well as complexity related to multiple usage of those procedures during the process based on, e.g., the o-GA which assigns the fitness $F(|B|, |C|)$ to each of its individuals, wherein F expresses the optimization goals specified in Definitions 12 and 14. The time complexity of Algorithm 2 (and Algorithm 1) equals to $O(n\,c(N))$ for $n = |A|$ and $N = |U|$, where $c(N)$ stands for the cost of verification of criteria of the form $_\vartheta B$ and $B \Rightarrow_\theta C$ within $(\Theta, \preceq, \Rightarrow_\theta)$. For example, in case of $\Rightarrow_\theta^H$ we may assume $c(N) = N \log N$, which is related to calculation of the quantities of $H(B)$ using formula (15) over $\mathbb{A} = (U, A)$ sorted by each given $B \subseteq A$ (cf. [29,40]). However, $c(N)$ may change given, e.g., specifics of various data types or implementations of sorting.

Algorithm 2 should be compared with other possibilities of applying rough set-based methods with respect to both results and complexity. Let us refer, e.g., to the ideas presented at the end of Section 3. If we want to adopt the technique of extracting $B \Rightarrow C$ based on identification of a given $B \subseteq A \setminus C$ as a decision reduct in the decision tables $\mathbb{A} = (U, A \setminus \{a\}, \{a\})$, $a \in C$, we should remember that approximate functional dependencies $B \Rightarrow_\theta \{a\}$ and $B \Rightarrow_\theta \{a'\}$ do not need to imply $B \Rightarrow_\theta \{c, c'\}$ for $\theta \neq \mathbf{0}$. As another example, when constructing an approximate association reduct from a previously found decision reduct $B \subseteq A \setminus \{a\}$ by adding attributes $a' \notin B \cup \{a\}$ to the right side of $(B, \{a\})$, we usually obtain the pairs (B, C) with both $|B|$ and $|C|$ lower than in Algorithm 2. From this perspective, the choice of a better procedure may depend on the choice of $F : \mathbb{N} \times \mathbb{N} \to \mathbb{R}$ in Definitions 12 and 14.

Algorithm 3. Modification of Algorithm 2

Input: $\theta, \vartheta \in \Theta$, $\theta \preceq \vartheta$; $\sigma : \{1, ..., 2n\} \to \{1, ..., 2n\}$

Output: $(B, C) \in 2^A \times 2^A$

  $B \leftarrow \emptyset$, $C \leftarrow \emptyset$

  for $i = 1$ to $2n$ **do**

    if $\sigma(i) \in \{1, ..., n\}$ **then**

      if $_\vartheta(B \cup \{a_{\sigma(i)}\})$ **then**

        $B \leftarrow B \cup \{a_{\sigma(i)}\}$

      end if

    end if

  end for

  for $i = 1$ to $2n$ **do**

    if $\sigma(i) \in \{1, ..., n\}$ **then**

      if $B \setminus \{a_{\sigma(i)}\} \Rightarrow_\theta C \cup \{a_{\sigma(i)}\}$ **then**

        if $_\vartheta(B \cup C \cup \{a_{\sigma(i)}\})$ **then**

          $B \leftarrow B \setminus \{a_{\sigma(i)}\}$, $C \leftarrow C \cup \{a_{\sigma(i)}\}$

        end if

      end if

    else

      if $B \setminus \{a_{\sigma(i)-n}\} \Rightarrow_\theta C$ **then**

        $B \leftarrow B \setminus \{a_{\sigma(i)-n}\}$

      end if

    end if

  end for

  return (B, C)

Algorithm 3 enables us to reach all (θ, ϑ)-association reducts. We pay the price of larger space of permutations and their longer processing time. However, if we rewrote Example 9 for Algorithm 3, we would notice that, again, the chance for a randomly chosen permutation to lead to a nearly optimal solution of $F\Theta_2$ or a component of a nearly optimal solution of $\Theta_2 \#_k$ should be relatively high.

Proposition 11. *Let* $\mathbb{A} = (U, A)$ *and* $(\Theta, \preceq, \Rightarrow_\theta)$ *satisfying (11,12) be given. Consider arbitrary* $\theta, \vartheta \in \Theta$, $\theta \preceq \vartheta$. *1) For every permutation* $\sigma : \{1, ..., 2n\} \to \{1, ..., 2n\}$, *the output of Algorithm 3 is a* (θ, ϑ)-*association reduct. 2) For every* (θ, ϑ)-*association reduct, there exists a permutation giving it as the output.*

Proof. The first part is similar to the proof of Proposition 10. In order to show the second part, consider a (θ, ϑ)-association reduct (X, Y) and $\sigma : \{1, ..., 2n\} \to \{1, ..., 2n\}$ such that, counting from $i = 1$, the first $|Y|$ cases such that $\sigma(i) \in \{1, ..., n\}$ correspond to the attributes $a_{\sigma(i)} \in Y$ and the next $|X|$ cases such that $\sigma(i) \in \{1, ..., n\}$ correspond to $a_{\sigma(i)} \in X$. Regardless of the cases of $\sigma(i) \in \{n+1, ..., 2n\}$, the first phase of Algorithm 3 results with $B \supseteq X \cup Y$ and $C = \emptyset$. Now, let us put $Z = B \setminus (X \cup Y)$. For $i = |Z| + 1, ..., |B|$, let $\sigma(i) \in \{1, ..., n\}$ correspond to the attributes in Y and X, as specified above. Additionally, if $Z \neq \emptyset$, let σ satisfy $a_{\sigma(i)-n} \in Z$ for each $\sigma(i) \in \{n + 1, ..., 2n\}$, $i = 1, ..., |Z|$. Then, at the end of the second phase we obtain $B = X$ and $C = Y$.

Algorithm 4. $(*,*)$-association reduct calculation for $(\Theta, \preceq, \Rrightarrow_\theta)$ and $\mathbb{A} = (U, A)$

Input: $\Theta' \subseteq \Theta$, $\Theta' = \{\theta_0, ..., \theta_m\}$; $\sigma : \{1, ..., m+n\} \to \{1, ..., m+n\}$
Output: $(B, C, \theta, \vartheta) \in 2^A \times 2^A \times \bigwedge^* \Theta' \times \bigwedge^* \Theta'$

  $B \leftarrow \emptyset,\ \vartheta \leftarrow \theta_0$
  for $i = 1$ **to** $m + n$ **do**
    if $\sigma(i) \in \{1, ..., m\}$ **then**
      if $\bigwedge\{\vartheta, \theta_{\sigma(i)}\} B$ **then**
        $\vartheta \leftarrow \bigwedge\{\vartheta, \theta_{\sigma(i)}\}$
      end if
    else
      if $\vartheta(B \cup \{a_{\sigma(i)-m}\})$ **then**
        $B \leftarrow B \cup \{a_{\sigma(i)-m}\}$
      end if
    end if
  end for
  $C \leftarrow \emptyset,\ \theta \leftarrow \vartheta$
  for $i = 1$ **to** $m + n$ **do**
    if $\sigma(i) \in \{1, ..., m\}$ **then**
      if $B \Rrightarrow_{\bigwedge\{\theta, \theta_{\sigma(i)}\}} C$ **then**
        $\theta \leftarrow \bigwedge\{\theta, \theta_{\sigma(i)}\}$
      end if
    else
      if $B \setminus \{a_{\sigma(i)-m}\} \Rrightarrow_\theta C$ **then**
        $B \leftarrow B \setminus \{a_{\sigma(i)-m}\}$
        if $B \Rrightarrow_\theta C \cup \{a_{\sigma(i)-m}\}$ **then**
          if $\vartheta(B \cup C \cup \{a_{\sigma(i)-m}\})$ **then**
            $C \leftarrow C \cup \{a_{\sigma(i)-m}\}$
          end if
        end if
      end if
    end if
  end for
  return $(B, C, \theta, \vartheta)$

Algorithm 4 refers to $(*,*)$-association reducts. It is the last considered example of a permutation-based approach to heuristic search for the solutions of optimization problems introduced in Section 8. It shows how to avoid too specific settings of thresholds $\theta, \vartheta \in \Theta$. The idea is to select a finite, reasonably small $\Theta' \subseteq \Theta$ and eventually restrict the analysis to the following subset of Θ:

$$\bigwedge\nolimits^* \Theta' = \{\bigwedge \Theta'' : \Theta'' \subseteq \Theta'\} \tag{26}$$

Proposition 12. *Consider* $\mathbb{A} = (U, A)$ *and* $(\Theta, \preceq, \Rrightarrow_\theta)$ *satisfying (11,12). Consider* $\Theta' = \{\theta_0, ..., \theta_m\}$ *and* $\sigma : \{1, ..., m+n\} \to \{1, ..., m+n\}$. *The output of Algorithm 4 is a* $(*,*)$-*association reduct for* $\mathbb{A}$ *when considered within the framework of* $(\bigwedge^* \Theta', \preceq, \Rrightarrow_\theta)$ *instead of* $(\Theta, \preceq, \Rrightarrow_\theta)$.

Proof. Straightforward extension of the proof of Proposition 10.

There are two interesting observations here: Firstly, by mixing attributes and approximation degrees within the same permutations, we can search for valuable $(*, *)$-dependencies with no rigid restriction to some particular thresholds. Secondly, consider, e.g., the entropy-based framework $([0, +\infty), \leq, \Rightarrow_\theta^H)$ and a finite $\Theta' \subseteq [0, H(A)]$. There is clearly $\bigwedge^* \Theta' = \Theta'$. When extending Example 8 for such a case, one would notice that the largest amounts of permutations lead to solutions $(B, C, \theta, \vartheta)$ with the values of θ and ϑ relatively closer to 0 than to $H(A)$, and with the pairs (B, C) relatively dissimilar from (θ', ϑ')-association reducts for the settings of (θ', ϑ') remaining relatively close to (θ, ϑ).

One can surely combine the ideas behind Algorithms 3 and 4 to obtain a more complete representation of $(*, *)$-association reducts, however, with even longer permutations to deal with. In its current form, Algorithm 4 has time complexity $O((n+m)c(N))$, wherein m may depend on N for some cases of $(\Theta, \preceq, \Rightarrow_\theta)$. As a result, we obtain a satisfactorily fast procedure that may be useful for searching for optimal solutions of such problems as $F^*\Theta_2^*$ or $\Theta_2 \#_k$, when rewritten for the framework of inference rules (11) and (12). As in the previous cases, such a procedure should be coupled with the appropriate mechanism of purely random or, e.g., o-GA-based heuristic search through the space of all permutations.

10   Conclusions

In this paper, we focused on extending the already well-established theoretical and algorithmic rough set-based foundations for information/decision reducts [25,26,29,41] onto a more advanced case of association reducts [30,31,32]. We compared association reducts with other methods of dealing with functional dependencies [2,4,11,37]. We discussed examples of approximate functional dependencies and their most important properties. We defined a number of optimization problems corresponding to various aspects of extracting such dependencies from data. In our opinion, further analysis of computational complexity of the formulated problems is one of the most challenging theoretical aspects in the area of the data-based approximate functional dependencies.

In Section 9, we extended the permutation-based algorithmic framework originally proposed for decision reducts in [40]. By modifying the input permutations in order to analyze the pairs of attribute sets at diversified levels or approximation, we obtained a methodology of searching for approximate association reducts. The proposed algorithms should be further compared with, e.g., modifications of the techniques searching for association rules [1,7] and relevant methods developed within the theory of rough sets [36,37]. Next steps should also include verification of the proposed framework over real-life data. In case of large data sets, one may consider, e.g., combining the introduced algorithms with the strategies of grouping attributes (cf. [13]) or objects (cf. [35]). Given the focus in applications evolving towards the ensembles of reducts, we will also continue our research on algorithms related to the optimization problems such as the one introduced in Definition 17. As illustrated in Section 9, the permutation-based mechanisms may be efficiently applied to such tasks in future.

References

1. Agrawal, R., Imieliński, T., Swami, A.N.: Mining Association Rules between Sets of Items in Large Databases. In: Proc. of SIGMOD 1993, Washington, DC, May 26–28, pp. 207–216 (1993)
2. Armstrong, W.W.: Dependency Structures of Database Relationships. Inform. Process. 74, 580–583 (1974)
3. Bazan, J.G., Nguyen, H.S., Nguyen, S.H., Synak, P., Wróblewski, J.: Rough Set Algorithms in Classification Problem. In: Rough Set Methods and Applications. New Developments in Knowledge Discovery in Information Systems. Studies in Fuzziness and Soft Computing, vol. 56, pp. 49–88. Physica Verlag (2000)
4. Bertet, K., Monjardet, B.: The multiple facets of the canonical direct unit implicational basis. Theoretical Computer Science (2009) (to appear)
5. Brown, E.M.: Boolean reasoning. Kluwer, Dordrecht (1990)
6. Burris, S.N., Sankappanavar, H.P.: A Course in Universal Algebra. Springer, Heidelberg (1981)
7. Ceglar, A., Roddick, J.F.: Association mining. ACM Comput. Surv. 38(2) (2006)
8. Davis, L. (ed.): Handbook of Genetic Algorithms. Van Nostrand Reinhold (1991)
9. Duentsch, I., Gediga, G.: Uncertainty Measures of Rough Set Prediction. Artif. Intell. 106(1), 109–137 (1998)
10. Gallager, R.G.: Information Theory and Reliable Communication. Wiley, Chichester (1968)
11. Ganter, B., Wille, R.: Formal Concept Analysis: Mathematical Foundations. Springer, Heidelberg (1998)
12. Garey, M.R., Johnson, D.S.: Computers and Intractability: A Guide to The Theory of NP-Completeness. Freeman and Company, New York (1979)
13. Grużdź, A., Ihnatowicz, A., Ślęzak, D.: Interactive Gene Clustering – A Case Study of Breast Cancer Microarray Data. Information Systems Frontiers 8(1), 21–27 (2006)
14. Hajek, P., Havranek, T.: Mechanizing Hypothesis Formation: Mathematical Foundations for a General Theory. Springer, Heidelberg (1978)
15. Kloesgen, W., Żytkow, J.M. (eds.): Handbook of Data Mining and Knowledge Discovery. Oxford University Press, Oxford (2002)
16. Kuncheva, L.I.: Combining Pattern Classifiers: Methods and Algorithms. Wiley, Chichester (2004)
17. Liu, H., Motoda, H. (eds.): Computational Methods of Feature Selection. Chapman & Hall, Boca Raton (2008)
18. McKinney, B.A., Reif, D.M., Ritchie, M.D., Moore, J.H.: Machine Learning for Detecting Gene-Gene Interactions: A Review. Applied Bioinformatics 5(2), 77–88 (2006)
19. Moshkov, M., Piliszczuk, M., Zielosko, B.: On Construction of Partial Reducts and Irreducible Partial Decision Rules. Fundam. Inform. 75(1–4), 357–374 (2007)
20. Nguyen, H.S.: Approximate Boolean Reasoning: Foundations and Applications in Data Mining. In: Peters, J.F., Skowron, A. (eds.) Transactions on Rough Sets V. LNCS, vol. 4100, pp. 334–506. Springer, Heidelberg (2006)
21. Nguyen, H.S., Nguyen, S.H.: Rough Sets and Association Rule Generation. Fundamenta Informaticae 40(4), 310–318 (1999)
22. Pawlak, Z.: Information systems theoretical foundations. Inf. Syst. 6(3), 205–218 (1981)

23. Pawlak, Z.: Rough sets – Theoretical aspects of reasoning about data. Kluwer, Dordrecht (1991)
24. Pawlak, Z.: Rough set elements. In: Rough Sets in Knowledge Discovery 1 – Methodology and Applications. Studies in Fuzziness and Soft Computing, vol. 18, pp. 10–30. Physica Verlag (1998)
25. Pawlak, Z., Skowron, A.: Rudiments of rough sets. Information Sciences 177(1), 3–27 (2007)
26. Skowron, A., Rauszer, C.: The discernibility matrices and functions in information systems. In: Intelligent Decision Support. Handbook of Applications and Advances of the Rough Set Theory, pp. 311–362. Kluwer, Dordrecht (1992)
27. Ślęzak, D.: Approximate reducts in decision tables. In: Proc. of IPMU 1996, Granada, Spain, July 1–5, vol. 3, pp. 1159–1164 (1996)
28. Ślęzak, D.: Various Approaches to Reasoning with Frequency Based Decision Reducts. In: Rough Set Methods and Applications. New Developments in Knowledge Discovery in Information Systems. Studies in Fuzziness and Soft Computing, vol. 56, pp. 235–288. Physica Verlag (2000)
29. Ślęzak, D.: Approximate Entropy Reducts. Fundamenta Informaticae 53(3–4), 365–390 (2002)
30. Ślęzak, D.: Association Reducts: A Framework for Mining Multi-Attribute Dependencies. In: Hacid, M.-S., Murray, N.V., Raś, Z.W., Tsumoto, S. (eds.) ISMIS 2005. LNCS, vol. 3488, pp. 354–363. Springer, Heidelberg (2005)
31. Ślęzak, D.: Association Reducts: Boolean Representation. In: Wang, G.-Y., Peters, J.F., Skowron, A., Yao, Y. (eds.) RSKT 2006. LNCS, vol. 4062, pp. 305–312. Springer, Heidelberg (2006)
32. Ślęzak, D.: Association Reducts: Complexity and Heuristics. In: Greco, S., Hata, Y., Hirano, S., Inuiguchi, M., Miyamoto, S., Nguyen, H.S., Słowiński, R. (eds.) RSCTC 2006. LNCS, vol. 4259, pp. 157–164. Springer, Heidelberg (2006)
33. Ślęzak, D.: Rough Sets and Few-Objects-Many-Attributes Problem – The Case Study of Analysis of Gene Expression Data Sets. In: Proc. of FBIT 2007, Jeju, Korea, October 11–13, pp. 437–440 (2007)
34. Ślęzak, D.: Degrees of conditional (in)dependence: A framework for approximate Bayesian networks and examples related to the rough set-based feature selection. Information Sciences 179(3), 197–209 (2009)
35. Ślęzak, D., Wróblewski, J., Eastwood, V., Synak, P.: Brighthouse: an analytic data warehouse for ad-hoc queries. PVLDB 1(2), 1337–1345 (2008)
36. Suraj, Z.: Discovery of Concurrent Data Models from Experimental Tables: A Rough Set Approach. Fundam. Inform. 28(3–4), 353–376, 379–490 (1996)
37. Suraj, Z.: Rough Set Method for Synthesis and Analysis of Concurrent Processes. In: Rough Set Methods and Applications. New Developments in Knowledge Discovery in Information Systems. Studies in Fuzziness and Soft Computing, vol. 56. Physica Verlag (2000)
38. Świniarski, R.W., Skowron, A.: Rough set methods in feature selection and recognition. Pattern Recognition Letters 24(6), 833–849 (2003)
39. Ullman, J.D., Garcia-Molina, H., Widom, J.: Database Systems: The Complete Book. Prentice Hall, Englewood Cliffs (2001)
40. Wróblewski, J.: Theoretical Foundations of Order-Based Genetic Algorithms. Fundamenta Informaticae 28(3–4), 423–430 (1996)
41. Wróblewski, J.: Ensembles of classifiers based on approximate reducts. Fundamenta Informaticae 47(3–4), 351–360 (2001)
42. Yao, Y.Y., Zhao, Y., Wang, J.: On Reduct Construction Algorithms. Transactions on Computational Science 2, 100–117 (2008)

Appendix A: Proof of Theorem 2 and Further Discussion

In [31], we formulated analogous result, but for the following Boolean formula instead of τ^*:

$$\wedge_{i,j:M_{ij}\neq\emptyset}((\vee_{a\in M_{ij}}(a\wedge a^*))\vee(\wedge_{a\in M_{ij}}a^*)) \tag{27}$$

Indeed, there are only two ways for a product term t to satisfy the part of the above formula indexed by a given $i, j = 1, ..., N$:

- To include terms a and a^* for at least one of attributes $a \in M_{ij}$ – it means that $a \in B$ (and $a \notin C$), i.e., $M_{ij} \cap B \neq \emptyset$ for (B, C) corresponding to t
- Otherwise, to include terms a^* for all attributes $a \in M_{ij}$, i.e., $M_{ij} \cap C = \emptyset$

Given such a relationship between satisfaction of formula (27) by t and satisfaction of discernibility-based criterion (5) by the corresponding (B, C), we refer to [31] – as well as to [26], where Boolean representation of information/decision reducts was first considered – for finalizing the proof. Here, it is enough to expose the only truly novel aspect, when comparing with [31]:

Lemma 1. *For each non-empty M_{ij}, there is the following logical equivalence:*

$$(\vee_{a\in M_{ij}}(a\wedge a^*))\vee(\wedge_{a\in M_{ij}}a^*)\equiv\wedge_{X:X\subseteq M_{ij},X\neq\emptyset}((\vee_{a\in M_{ij}\setminus X}a)\vee(\vee_{a\in X}a^*)) \tag{28}$$

Proof. Straightforward application of Boolean distributive/absorption laws [5].

Let us finish this section with continuing the case study of $\mathbb{A} = (U, A)$ in Table 1. Given the contents of $\mathbb{M}$, a "sample" of the formula (27) looks as follows:

$$(aa^* \vee bb^* \vee cc^* \vee a^*b^*c^*)(bb^* \vee ee^* \vee b^*e^*)...(aa^* \vee dd^* \vee ff^* \vee a^*d^*f^*)$$

Further, it can be transformed to formula (8) in CNF:

$$\tau^* \equiv (a \vee b \vee c^*)(a \vee b^* \vee c)(a \vee b^* \vee c^*)(a^* \vee b \vee c)(a^* \vee b \vee c^*)(a^* \vee b^* \vee c)$$
$$(a^* \vee b^* \vee c^*)(b \vee e^*)(b^* \vee e)(b^* \vee e^*)...(a \vee d \vee f^*)(a \vee d^* \vee f)$$
$$(a \vee d^* \vee f^*)(a^* \vee d \vee f)(a^* \vee d \vee f^*)(a^* \vee d^* \vee f)(a^* \vee d^* \vee f^*)$$

Application of absorption laws to simplify formulas (27) and (8) is not so straightforward as in case of information and decision reducts. Still, some simple tricks are possible. For example, a single-attribute element of $\mathbb{M}$, like e.g. $M_{16} = \{f\}$, corresponds to a trivial clause f^* in τ^*, which enables to immediately remove any other clauses containing f^*, e.g.: $(a \vee d \vee f^*)$, $(a \vee d^* \vee f^*)$, $(a^* \vee d \vee f^*)$ and $(a^* \vee d^* \vee f^*)$. For more detailed studies on shortening the Boolean formulas of the form (27), we refer to [31].

Let us also note that the size of τ^* in CNF is far larger than that of Boolean formulas representing information and decision reducts. Indeed, in case of association reducts, the number of clauses is exponential with respect to the cardinality of A (though it is still quadratic with respect to the cardinality of U), with no straightforward ability to push it down to any polynomial bounds. It supports an intuitive expectation that the families of association reducts derivable from data can be far larger than those of information/decision reducts. It certainly requires further theoretical and algorithmic investigations.

Appendix B: Proof of Theorem 4

We proceed like in the proof of Theorem 3 [32]. We reduce the Minimal Dominating Set Problem (MDSP) to $F\Theta_2$. MDSP, known as NP-hard [12], is defined by INPUT as undirected graph $\mathcal{G} = (A, E)$ and OUTPUT as the smallest $B \subseteq A$ such that $Cov_{\mathcal{G}}(B) = A$, where $Cov_{\mathcal{G}}(B) = B \cup \{a \in A : \exists_{b \in B}(a, b) \in E\}$. To reduce MDSP to $F\Theta2$, we construct information system $\mathbb{A}_{\mathcal{G}} = (U_{\mathcal{G}}, A_{\mathcal{G}})$, $U_{\mathcal{G}} = \{u_1, ..., u_n, o_1, ..., o_n, u_*\}$, $A_{\mathcal{G}} = \{a_1, ..., a_n, a_*\}$, $n = |A|$, as follows:

$$a_i(u_j) = 1 \Leftrightarrow i = j \vee (i, j) \in E \qquad a_i(u_j) = 0 \text{ otherwise}$$
$$a_i(o_j) = 1 \Leftrightarrow i = j \qquad\qquad a_i(o_j) = 2 \text{ otherwise} \qquad (29)$$
$$a_i(u_*) = 0, \qquad a_*(u_j) = 0 \qquad a_*(o_j) = 0, \qquad a_*(u_*) = 1$$

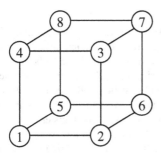

Fig. 1. $\mathcal{G} = (A, E)$ with 8 nodes and $\mathbb{A}_{\mathcal{G}} = (U_{\mathcal{G}}, A_{\mathcal{G}})$ constructed using (29)

$U_{\mathcal{G}}$	a_1	a_2	a_3	a_4	a_5	a_6	a_7	a_8	a_*
u_1	1	1	0	1	1	0	0	0	0
u_2	1	1	1	0	0	1	0	0	0
u_3	0	1	1	1	0	0	1	0	0
u_4	1	0	1	1	0	0	0	1	0
u_5	1	0	0	0	1	1	0	1	0
u_6	0	1	0	0	1	1	1	0	0
u_7	0	0	1	0	0	1	1	1	0
u_8	0	0	0	1	1	0	1	1	0
o_1	1	2	2	2	2	2	2	2	0
⋮									⋮
o_8	2	2	2	2	2	2	2	1	0
u_*	0	0	0	0	0	0	0	0	1

Lemma 2. *[32] For $B \subseteq \{a_1, ..., a_n\}$, $B \Rightarrow \{a_*\}$ holds in $\mathbb{A}_{\mathcal{G}}$, iff $Cov_{\mathcal{G}}(B) = A$.*

Lemma 3. *[32] For $X, Y \subseteq A_{\mathcal{G}}$, $X \cap Y = \emptyset$, (X, Y) is an association reduct in $\mathbb{A}_{\mathcal{G}}$, iff $X = Y = \emptyset$ or $Y = \{a_*\}$ and X is an irreducible dominating set in $\mathcal{G}$, i.e. $Cov_{\mathcal{G}}(X) = A$ and $Cov_{\mathcal{G}}(X') \neq A$ for every $X' \subsetneqq X$.[7]*

Now, to complete the proof of Theorem 4, the following observations are enough:

1. Each F satisfying (23), after disregarding $(\emptyset, \emptyset)$, reaches its maximum for the smallest dominating sets – this is because the lower n, the higher $F(n, 1)$
2. For a given $\mathbb{A}_{\mathcal{G}} = (U_{\mathcal{G}}, A_{\mathcal{G}})$, we can put $\theta = \mathbf{0}$ (14), as well as $\vartheta \in \Theta$ such that $_{\vartheta}A_{\mathcal{G}}$ holds, as input to $F\Theta2$ – then the solution (B, C) of $F\Theta2$ for $\mathbb{A}_{\mathcal{G}}$ yields B as the solution of MDSP for $\mathcal{G}$

The setting of ϑ is the only difference with respect to the proof of Theorem 3 in [32]. It is an easy observation that by considering $\vartheta \in \Theta$ such that $_{\vartheta}A_{\mathcal{G}}$ is satisfied (so we have also $_{\vartheta}X$ for any $X \subseteq A_{\mathcal{G}}$), we actually draw equivalence between θ-association and (θ, ϑ)-association reducts in $\mathbb{A}_{\mathcal{G}}$. (See also Example 9.) This is why the rest of construction can be the same as in [32] indeed.

---

[7] Note that for every graph there is always at least one non-empty dominating set.

Hybrid Evolutionary Algorithm for the Graph Coloring Register Allocation Problem for Embedded Systems

Anjali Mahajan[1] and M.S. Ali[2]

[1] G H Raisoni College of Engineering
Nagpur, India
armahajan@rediffmail.com
[2] Prof. Ram Meghe Institute of Technology and Research,
Badnera, Amravati, India
softalis@hotmail.com

Abstract. Memory or registers are used to store the results of computation of a program. As compared to memory, accessing a register is much faster, but they are scarce resources, in real-time embedded systems and have to be utilized very efficiently. If the register set is not sufficient to hold all program variables, certain values have to be stored in memory and so-called spill code has to be inserted. The optimization goal is to hold as many live variables as possible in registers in order to avoid expensive memory accesses. The register allocation phase is generally more challenging in embedded systems In this paper, we present a new hybrid evolutionary algorithm (HEA) for graph coloring register allocation problem for embedded systems based on a new crossover operator called crossover by conflict-free sets (CCS) and a new local search function. The objective is to minimize the total spill cost.

1 Introduction

Register allocation is a fundamental compiler component that allocates and assigns variables and temporaries to registers. It is one of the most important optimizations a compiler performs and is becoming increasingly important as the gap between processor speed and memory access time widens. Its goal is to find a way to map the temporary variables used in a program into physical memory locations (either main memory or machine registers). Accessing a register is much faster than accessing memory, therefore one tries to use registers as much as possible. Of course, this is not always possible, thus some variables must be transferred (spilled) to and from memory. This has a cost, the cost of *load* and *store* operations, which should be avoided as much as possible. Typically, this degrades runtime performance and increases power consumption. Therefore, an efficient mapping should generally minimize the register requirements and the number of spilling instructions. The critical applications in embedded computing, industrial compilers are ready to accept longer compilation times if the final code gets improved.

Although graph-coloring is very suitable for RISC-like architectures. With plenty of registers it cannot be simply adopted for irregular architectures. The extension for

M.L. Gavrilova et al. (Eds.): Trans. on Comput. Sci. V, LNCS 5540, pp. 206–219, 2009.

graph-coloring were proposed in [12] [1] which fail to give a unified solution for irregular architectures. The problems like hardware restrictions modeling, irregularities increase the complexity of the interference graph which badly affects graph coloring.

Graph coloring abstracts the problem of assigning registers to live ranges in a program into the problem of assigning colors to nodes in an *interference graph*. The register allocator attempts to "color" the graph with a finite number of machine registers. The only constraint is that any two nodes connected by an interference edge must be colored with different registers.

The compiler first constructs an interference graph G to model register allocation as a graph coloring problem. The nodes in G correspond to live ranges, and the edges represent interferences. Thus, there is an edge in G from node i to node j if live range of i interferes with live range of j, that is, if they are simultaneously live at some point and cannot occupy the same register.

To find an allocation from G, the compiler looks for a k-coloring of G. It assigns k colors to the nodes of G such that adjacent nodes always have distinct colors. If we choose k to match the number of machine registers, then we can map a k-coloring of G into a feasible register assignment for the underlying code. As graph coloring is NP-complete[3], the compiler uses a heuristic method to search for a coloring. It is not guaranteed to find a k-coloring for all k-colorable graphs. If a k-coloring is not found, some values are spilled. Spilling means, the values are kept in memory rather than in registers.

Spilling one or more live ranges creates a new and different interference graph. The compiler proceeds by iteratively spilling some live ranges and attempting to color the resulting new graph.

In the context of embedded processors, the most important restriction of the graph coloring approach is that it is based on the assumption of a homogenous register set. The different phases of register allocation are differently impacted by embedded processor irregularities.

This paper presents a heuristic algorithm for graph-coloring register allocation problem for embedded systems based on a new crossover operator called crossover by conflict-free sets (CCS) and a new local search function.

2 Problem Description

The register allocation problem is briefly presented in this section. Assume that $V=\{v_1,v_2,v_3,...\}$ is the set of variables in a given intermediate representation of a program. A variable $v_i \in V$ is said to be *live* at a given point in the program if it is defined above and has not been used yet for the last time. Values stay in registers as long as they are live. Live range LR_i for a variable v_i is the region which begins with the definition of vi and ends with the last use of v_i.

If any two live ranges LR_i and LR_j are simultaneously live at some point p in the program flow, they cannot be stored in the same architectural register. In this case, we say that LR_i interferes with LR_j . To model the allocation problem, a compiler constructs an *interference graph*, $IG=(V,E)$, where V is the set of individual live ranges and E is the set of edges that represent interferences between live ranges.

A variable may have more than one live range and a register allocator can, and in most cases will, keep these live ranges in different physical registers. To implement this, if an instruction has the same register name in both destination and source fields, a different register name can be used in the destination field in order to provide a new live range.

The register allocation problem is formally defined as the problem of mapping f: $V \rightarrow R = \{r_1, r_2, .. r_k\}$, where R is the set that holds a maximum of k registers such that if $(v_i, v_j) \in E$, then $f(v_i) \neq f(v_j)$. When the interference graph is constructed, the compiler looks for an assignment of k colors to the nodes (vertices) of IG, with distinct colors for adjacent nodes. This process corresponds to the well-known graph coloring problem of finding a k-coloring of the interference graph.

3 Related Work

Global register allocation has been studied extensively in the literature. The predominant approach, first proposed by Chaitin et al. [4], implements the graph-coloring register allocator, the heuristic technique which can usually be used to do the coloring quickly in practice. Register allocation abstracts the problem of assigning CPU registers to symbolic registers (variables) into problem of coloring vertices in an interference graph [4]. The vertices of the interference graph represent symbolic registers and edges of the graph imply interference constraints (two symbolic registers connected by an edge cannot be assigned the same CPU register). A coloring of this graph is equivalent to an assignment of CPU registers.

This approach considers plenty of general purpose registers and an orthogonal instruction set. However, it is very difficult to adopt the graph coloring approach for irregular architectures since it assumes that a vertex can be colored with any color as long as the interference constraints hold. This assumption is wrong for irregular architectures.

It is no doubt that the original graph coloring register allocation algorithm [4] is a success, but that does not mean it is the only algorithm in register allocation.

Briggs [2] introduced an improved coloring strategy that produces better allocations for many graphs on which Chaitin's method fails. The key difference between Briggs' algorithm and Chaitin's algorithm lies in the timing of spill decisions. By optimistically assuming that nodes of high degree will receive colors, Briggs' method inserts a spill candidate into the coloring order. The spill candidate either receives a color or it does not in which case the allocator spills it.

Work on graph coloring [2] [6] has focused on removing unnecessary moves in a conservative manner so as to avoid introducing spills. Briggs also describes an algorithm for coloring aligned and unaligned register pairs [1]. This algorithm requires that a node's degree accurately reflect its colorability. To make a node's degree reflect its colorability even in the presence of aliasing, Briggs adds "additional" edges to the interference graph in an attempt to model the aliasing constraints.

Briggs et al. [2] describe two improvements to Chaitin-style graph coloring register allocators. The first, optimistic coloring uses a stronger heuristic to find a k-coloring for the interference graph. The second extends Chaitin's treatment of rematerialization to handle a larger class of values.

An approach for combinatorial optimization is to embed local search into the framework of population based evolutionary algorithm, leading to hybrid evolutionary algorithm. The crossover operator is the most important element to be considered while designing a Hybrid Evolutionary Algorithm.

Topcuoglu et al. in [13] propose a highly specialized crossover operator called conflict-free partition crossover (CFPX) that incorporates the specific information on the register allocation problem. The CFPX operator aims to providing conflict-free register classes successively, by constructing a single class of offspring at each step. They propose an efficient local search method for checking the neighborhood solutions, where the search area decreases with the decrease in the number of conflicts in the offspring. The CFPX operator combined with a problem-specific local search leads to a very powerful method, which places this approach into the class of the Hybrid Evolutionary Algorithms.

To find suboptimal solutions of high quality for the GCP, [5] developed powerful hybrid evolutionary algorithms. To achieve this, Galinier and Hao devised a new class of highly specialized crossover operators. These crossovers, combined with a well-known tabu search algorithm, lead to a class of Hybrid Evolutionary Algorithms. They choose to experiment a particular Hybrid Evolutionary Algorithm. Results on large benchmark graphs prove very competitive with and even better than those of state-of-the-art algorithms.

4   A Hybrid Evolutionary Algorithm

Evolutionary algorithms involve natural evolution and genetics. The genetic algorithm is a classical method in this category. It has been applied to various optimization problems. There are several other methods like genetic programming, ant colony optimization, etc. The simple evolutionary algorithms are not generally efficient for complex combinatorial problems. The performance of the evolutionary algorithms is improved with the addition of problem specific knowledge. Specialized operators are combined with evolutionary algorithms to generate complex hybrid systems called hybrid genetic algorithms, hybrid evolutionary algorithms, genetic local search algorithms and memetic algorithms.

An approach for combinatorial optimization is to embed local search into the framework of population based evolutionary algorithm, leading to hybrid evolutionary algorithm. HEA is based on two elements: an efficient local search (LS) operator and a highly specialized crossover operator. The basic idea consists in using the crossover operator to create new and potentially interesting configurations which are then improved by the LS operator.

We present a hybrid evolutionary algorithm for graph coloring based on a new crossover operator for register allocation problem and a new local search function. This section presents our algorithm- Hybrid evolutionary Algorithm for Graph coloring Register Allocation with proposed operator called crossover by conflict-free sets(CCS).

4.1   The Algorithm

This section presents the proposed algorithm- Hybrid evolutionary Algorithm for Graph coloring Register Allocation (HEA).

It consists of classic graph coloring algorithm based on Chaitin [4] followed by our algorithm. If the interference graph is simple enough, then the former can color it successfully, and the latter does nothing. Otherwise, the former is used to make the interference graph simpler and color insignificant nodes, and the remaining complex graph is given to the latter. Chaitin algorithm is used to remove unnecessary move instructions by *aggressive_coalesce()* subroutine. Then it calls *simplify()* routine that repeatedly removes nodes having fewer than k neighbors and their associated edges from the interference graph and pushes these nodes on a stack. At some point, either we obtain the empty graph, in which case we can produce a proper k-coloring for the original graph by popping each node from the stack and giving it a color distinct from its neighbors, or we obtain a graph whose nodes are all high-degree(having k or more neighbors). In the latter case, this new high-degree interference graph is given to HEA for assigning variables to register classes.

If a k-coloring is not discovered, some values are spilled; that is, the values are kept in memory rather than in registers. The spill cost of such variables is calculated. It is one of the metrics used in performance evaluation.

The HEA consists of a genetic component and a local search (LS). The genetic component initializes and evolves a population of solutions. The general algorithm is as given below:

Input : *Interference graph , IG = (V,E) ; number of registers , k*
Output : *best configuration*
begin
    P = generate_population(|P|)
    iter = 0
    while (iter $\leq$ MaxIter **or** popu-diversity>0) **do**
        (p1, p2) = select_parents(P)
        p = crossover (p1, p2)
        p = local_search(p, L)
        P = update_population(P,p)
        iter =iter + 1
    endwhile
end

The algorithm first builds an initial population of configurations (*generate _ population)* and then performs a series of cycles called generation. At each generation, two configurations *p1* and *p2* are chosen in the population (*select_parents)*. A crossover is then used to produce an offspring *p* from *p1* and *p2* (*crossover)*. The LS operator is applied to improve *p* for a fixed number L of iterations (*local_search)*. Finally, the improved configuration p is inserted in the population by replacing another configuration (*update_population)*. This process repeats until the value of *iter* is less than or equal to a prefixed number *MaxIter* is reached or population diversity (*popu-diversity)* is greater than zero.

Population diversity is calculated as the average distance between all configurations in the population. For two configurations p1 and p2, the distance between p1 and p2 is the minimum number of elementary transformations necessary to transform p1 into p2.

In our approach, we consider the partition method for string representation [5]. Each solution P_i partitions the variables into register classes, $P_i = \{R_1, R_2, \ldots R_k\}$ where each class R_i includes the live ranges of variables that are mapped to the registers r_i and k is the total number of registers. Given two parents p1= $\{R^1_1, R^1_2, \ldots R^1_k\}$ and p2= $\{R^2_1, R^2_2, \ldots R^2_k\}$, the partial configuration will be a set $\{R_1, R_2, \ldots R_k\}$ of disjoint sets of nodes where each subset R_i is included in a class of one of the two parents, and all R_i are conflict-free sets. (the nodes i and j in an interference graph are said to be conflict-free, when there is no edge connecting them)

4.2 Initial Population Generation

Generally the initial population is generated using the DSatur algorithm [3], which is a graph coloring heuristic. It considers the *saturation degree* of each node, which can be defined as the number of different colors to which the node is adjacent to. In the register allocation problem, both the spill costs and the degree of the nodes in a given interference graph are considered. We adopt the new metric called the *spill degree* proposed by [13] that can be used for ordering the nodes. The spill degree of a node i can be defined by one of the three equations given below

$$SDegree_1(i) = SCost(i) \times Degree(i)$$
$$SDegree_2(i) = SCost(i) \times Degree^2(i)$$
$$SDegree_3(i) = SCost(i) \tag{1}$$

In these expressions, *SCost(i)* is the spill cost and *Degree(i)* is the number of edges incident to node i. In order to generate the initial population, the spill degrees are set using a combination of the three equations given in (1). The nodes are sorted in decreasing order of spill degrees. At each iteration, an unsigned node with the maximum spill degree is mapped to the register class with the lowest possible index value, where the selected node should be conflict free with the other nodes in the same class. This process is repeated until all the nodes (i.e., their corresponding variables) are mapped to one of the register classes.

$$RCF\ (i) = \sum_{\forall m \in Ri} Conflict\ (m) \tag{2}$$

4.3 The Crossover Operator

The crossover used here is the new proposed operator Crossover by Conflict-free Sets (CCS). Given two parent configurations p1= $\{R^1_1, R^1_2, \ldots R^1_k\}$ and p2= $\{R^2_1, R^2_2, \ldots R^2_k\}$, chosen randomly by the *select_parents* operator from the population, the algorithm *crossover (p1, p2)* builds an offspring p= $\{R1, R_2, \ldots R_k\}$ as follows

Input: *configurations p1= { $R^1_1, R^1_2, \ldots R^1_k$}*
     and p2= { $R^2_1, R^2_2, \ldots R^2_k$}
Output: *configuration p= { $R_1, R_2, \ldots R_k$}*
begin
// consider p1
if *conflict(p1)>0* ***then***
                call function *conflict_free(p1)*

// **consider p2**
if conflict(p2)>0 then
                        call function conflict_free(p2)
    for i (1≤ i ≤ k) do

CfR^1_n= max $_{q∈p1}$ Cf_Individual(q)
// CfR^1_n is the partition with maximum number of conflict-free variables from p1
CfR^2_n= max $_{q∈p2}$ Cf_Individual(q)
// CfR^2_n is the partition with maximum number of conflict-free variables from p2
            choose j such that | CfR^1_n∩ CfR^2_j| *is maximum*
            R_i = |CfR^1_n ∪ CfR^2_j|
            SpillQuality(R_i)=Spillcost(R_i)*Reg_class_Conflict(R_i)
            remove the nodes of R_i from p1 and p2

    endfor
    assign randomly the nodes of R- (R_l U...U R_k)
end

function *conflict_free(p)*
begin
 Cf_Individual=∅
for i (1≤ i ≤ k) do
    if Reg_class_Conflict(i) >0 then
// Reg_class_Conflict(i) is the total number of conflicts in i^{th} register class
begin
initialize t to 0
while *(t < (Reg_class_Conflict(i)/2)) do*
            select a variable m from R_i where
            Conflict(m)=max$_{c∈Ri}${Conflict(c)} or
            spillcost(m)=min$_{c∈Ri}${spillcost(c)}
            remove variable m from R_i
            t=t+Conflict(m)
endwhile
CfR(i)= R_i
endif
Cf_Individual= Cf_Individual∪ CfR(i)
  endfor
end

The algorithm builds step by step the k classes R_1 , R_2 ... R_k of the offspring. If any register set of the parent has conflict, the algorithm first determines the conflict free sets of all the register classes of each parent. The function to find the conflict free set is based on the heuristic from [13] for determining the conflict-free set with the maximum number of nodes of each register class. In that, initially the conflict-free set includes all variables in R^l. The total number of conflicts of each variable m in class R^l, conflict(m) are determined. The sum of the *conflict(m)* values gives us the total number of conflicts in the register class R^l, *Reg_class_Conflict(l)*. Then, the variables from the partition are removed one by one in decreasing order of *conflict(m)* values, until the total number of conflicts in R^l becomes equal to ½ of its initial value. The function gives us the conflict free register classes of each individual parent. We then consider the partition $CfR^l{}_n$, which has *maximum number of conflict-free nodes from p1*. It then selects a partition $CfR^2{}_j$ from p2 that has largest number of nodes in common with $CfR^l{}_n$. The algorithm unifies the pair of these two partitions. It calculates the conflict factor of the produced partition. If it is zero, the partition is conflict free. It then removes all the nodes belonging to offspring register partition from the parents p1 and p2. In case of tie-breaking, any partition is taken randomly. This new partition becomes the base set of the first register class for the offspring. The process is subsequently repeated for the other register classes.

At the end of k steps, some nodes may remain unassigned. These are then assigned to a class which has minimum *conflict factor*.

4.4 The Local Search LS Operator

After a solution is generated using the crossover operator, the local search phase improves it before inserting it into the population for a maximum of L iterations. We have used a Local search operator LS. It uses a 1-exchange neighborhood. Formally, given a partition $\{R_1 , R_2 , \dots R_k\}$, a neighboring partition is obtained by changing the color of one node, i.e., a node v_i is removed from the class R_i to which it belongs, and it is moved into a class R_j. The evaluation function has two components: the first one favors large color classes, while the second one reduces the number of conflicts. If $E(R_i)$ is the set of edges from E that have both ends in R_i, the evaluation function is given in (3).

Local minima for this function corresponds to configuration p , created by crossover and improved by LS to be the conflict free assignment of variables to register classes. It is now inserted in the population by replacing the worst of the two parents.

$$f = -\sum_{i=1}^{k} |R_i|^2 + \sum_{i=1}^{k} 2 |R_i| \| E_i |$$

$$3)$$

4.5 Fitness Function Calculation

To solve a register allocation problem, we consider a set of k - register classes. In the interference graph IG = (V, E), all variables (i.e., nodes) v_i are assigned to k - register classes. The nodes i and j in an interference graph are said to be conflict-free, when there is no edge connecting them. *Conflict(m)* denotes the conflict of a variable m in register class R_i. *Spillcost* is the spill cost of variable m. The fitness of a register class is given as

$$fitness(R_i) = \sum_{\forall m \in Ri} Conflict(m)*Spillcost(m)/degree(m) \qquad (4)$$

The total sum of fitness values of all register classes gives the fitness value of the given individual

$$Fitness = \sum_{i=1}^{k} fitness(i) \qquad (5)$$

The goal of the optimization process is to minimize fitness until zero.

5 Experimental Evaluation

The experiments are based on MachineSUIF [14] compiler research framework.

A widely-used algorithm is iterated register coalescing proposed by Appel and George [6], a modified version of previous developments by Chaitin [4], and Briggs et al. [1] [2]. In these heuristics, *spilling*, *coalescing* (removing register-to-register *moves*), and *coloring* (assigning a variable to a register) are done in the same framework.

We compare the results with the best results published so far which are those obtained by the GPX based allocator using Hybrid Evolutionary Algorithm of Galinier and Hao [5], Chaitin allocator [4], Briggs allocator [1] [2] and George-Appel's 'Iterated Register Coalescing' algorithm [6].

The register allocator of MachineSUIF implements George-Appel's 'Iterated Register Coalescing' algorithm. We have implemented Chaitin allocator, Briggs allocator, GPX based allocator and our own HEA allocator. We replace the existing original allocator in SUIF/MachineSUIF compiler research framework by these allocators; other parts of the framework are not changed.

In order to evaluate the effectiveness of the proposed algorithm, it is compared with all these algorithms[9][10].

5.1 Benchmarks

We applied the algorithms to 10 embedded applications selected from Mediabench [8], MiBench[7] and 5 real time applications from SNU-RT [15]. For each application, the population size is set to 50 and LS iterations are all set to 200. For each application, we run the allocators five times and average the results.

Table 1. Number of registers used

Benchmark	Chaitin	George-Appel	Briggs	GPX based	HEA
pegwit	20	15	16	15	14
jpeg_cjpeg	28	22	23	21	20
crc32	14	10	10	9	8
dijkstra	9	6	6	6	6
gsm	19	17	18	17	16
patricia	14	10	11	10	10
mpeg2dec	26	21	22	21	20
sha	15	12	13	11	11
FFT	19	15	15	15	14
blowfish	18	14	15	14	14
bs	9	7	7	6	6
fftl	15	12	12	11	12
matmul	21	18	18	17	17
qsort_exam	23	14	15	13	12
select	19	16	16	13	12

5.2 Number of Registers Used

Table 1 lists each of the benchmark and the number of registers used in the respective experiment. For each benchmark the number of registers allocated was the maximum chromatic number among the interference graph of each procedure.

Of all these algorithms, the Hybrid Evolutionary Algorithms based algorithms gives better results. It is found that in many cases, GPX based algorithm and our HEA are comparable. It is due to the Tabu search meta heuristics used in [5]. From the three traditional algorithms, it is seen that the performance of George –Apple is better.

5.3 Spill Loads

Spill loads predict whether a method will benefit from the additional effort of applying the more expensive graph coloring allocator. Before looking at execution times on an actual machine, we consider the quality of the induced hybrid allocator (compared with always applying either graph coloring or linear scan) in terms of the number of spill loads added by register allocation.

The spills cause loads and stores which results in a higher execution time and the register allocator has to decide which symbolic register is cheaper to spill.

Table 2 gives the spill loads of all the algorithms. As shown in table 3, the spill load of Chaitin algorithm is far more than that of other algorithms. The spill loads produced by hybrid evolutionary algorithms is minimum. However, the benchmark FFT does not perform well in the heuristic of HEA because a bad local spill-decision was taken due to the heuristic.

Table 2. Spill load of Instructions

Benchmark	Chaitin	George -Appel	Briggs	GPX based	HEA
pegwit	89	76	78	47	49
jpeg_cjpeg	459	282	289	240	256
crc32	345	135	167	89	78
dijkstra	720	412	401	345	358
gsm	328	252	276	198	202
patricia	254	124	130	94	89
mpeg2dec	88	43	52	42	45
sha	219	177	180	164	167
FFT	315	288	294	265	268
blowfish	797	367	543	87	89
bs	59	4	5	0	0
fftl	289	67	70	41	43
matmul	156	45	48	14	15
qsort_exam	234	12	14	0	0
select	453	26	28	13	12

It is due to the aggressive coalescing strategy of Chaitin algorithm that changes many nodes of interference graph from low-degree to high-degree, which makes the condition of 'degree<k' unsatisfied. So it has to select a node, which may be unnecessary. Therefore the performance of Chaitin algorithm is far worse than that of the other algorithms. Genetic operator systematically eliminates low-quality solutions from the population, preserve diversity between solutions, and provide better input for local search. A small amount of spill cost is due to function callers and callees saving many contents of registers in order to preserve correct program semantics.

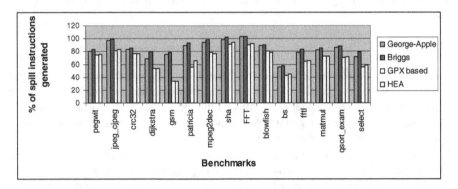

Fig. 1. Percentage of Spill Instructions Generated with respect to George-Appel's Allocator

Table 3. Total execution time

| Benchmark | Execution Time | | | Improvement % | |
	George-Appel	GPX based	HEA	GPX based - George-Appel	HEA - George-Appel
pegwit	1675892	1546237	1534732	7.74	8.42
jpeg_cjpeg	8985	8165	8123	9.13	9.59
crc32	234167	225671	225671	3.63	3.63
dijkstra	763542	745232	742536	2.40	2.75
gsm	833446	823456	827654	1.20	0.69
patricia	9486463	9351235	9352781	1.43	1.41
mpeg2dec	6665775	6371282	6371282	4.42	4.42
sha	438927	427861	423567	2.52	3.50
FFT	6465692	6465234	6465834	0.01	0.01
blowfish	8903126	8768952	8962342	1.51	-0.67
bs	24732762	23456752	23456752	5.16	5.16
fftl	84227	83452	83456	0.92	0.92
matmul	5637653	5634523	5634523	0.06	0.06
qsort_exam	1237865	1236872	1236754	0.08	0.09
select	243645	233622	233619	4.11	4.12

Fig. 1 shows the percentage of the spill instructions generated by each allocator to that of spill instructions generated by George-Appel's allocator. The graph shows that GPX based and HGR allocators generate a small number of spill instructions than the classical allocators.

Table 3 shows the total execution time of the benchmarks. In the first improvement column (GPX based - George-Appel) we compare the execution time by using the optimal GPX based algorithm with the execution time by using the George-Appel algorithm. As given in the table the improvements range from 0.01 % to 9.59 %. The improvement by using the HEA algorithm (HEA - George-Appel) still performs significantly better than the graph coloring except for the benchmarks gsm, patricia. For the benchmark blowfish the HEA algorithm performs poorly.

Fig. 2 shows the % improvement in execution time over the baseline allocator which is George-Appel's allocator. Our experiments demonstrate that our HEA approach gives significantly better results than a traditional graph coloring approach like George- Appel's allocator. They are comparable to those of GPX based allocator, except for the benchmark blowfish.

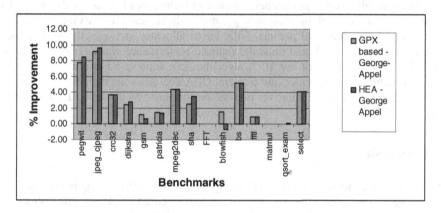

Fig. 2. Percentage improvement in execution time

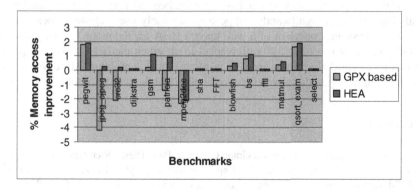

Fig. 3. Improvement of memory accesses

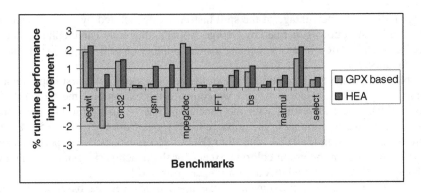

Fig. 4. Runtime performance

Fig. 3 and Fig. 4 show the improvement of memory accesses and runtime performance of HEA and GPX based allocator over the baseline.

GPX based allocator degraded performance in some cases because of the increased register pressure caused by eliminating some redundancies. For example, for the benchmarks jpeg_cjpeg and patricia, GPX based allocator increases the memory accesses by 4.2% and 1.4% and thus, the runtime performance was degraded by 2.1% and 1.5%. However, use of our HEA algorithm can achieve a performance benefit. For the benchmarks jpeg_cjpeg and patricia, HEA decreases the memory accesses by 0.27% and 0.9%, and thus, improves runtime performance by 0.7% and 1.2% over the baseline.

6   Conclusion

In this paper, we introduced a new hybrid algorithm for graph coloring register allocation using genetic algorithm and local search. Compared to classic graph coloring register allocation algorithms, our algorithm can deal with irregular architectural characteristics of embedded processors more uniformly and can generate better object code. The experimental evaluation based on embedded and real-time benchmarks reveals that our HEA significantly out perform a widely used register allocation heuristic and a crossover operator of a well known HEA for the graph coloring problem (the GPX operator given in [5]), with respect to all given metrics of solution quality. It proves that hybrid algorithms are very powerful for this problem. Moreover, a HEA like ours is not computationally expensive to obtain high quality solution.

References

1. Briggs. P.: Register Allocation via Graph Coloring. Ph.d Thesis, Rice University (1992)
2. Briggs, P., Cooper, K., Torczon, L.: Improvements to graph coloring register allocation. ACM Transactions on Programming Languages and Systems 16(3), 428–455 (1994)
3. Brelaz, D.: New methods to color the vertices of a graph. ACM Communication 22(4), 251–256 (1979)

4. Chaitin, G.: Register Allocation and Spilling via Graph Coloring. In: ACM SIGPLAN 1982 Symposium on Compiler Construction. SIGPLAN Notices, vol. 17(6), pp. 98–105. ACM Press, New York (1982)
5. Galinier, P., Hao, J.: Hybrid evolutionary algorithms for graph coloring. J. of Combinatorial Optimization 3, 379–397 (1999)
6. George, L., Appel, A.: Iterated Register Coalescing. ACM Transactions on Programming Languages and Systems 18(3), 300–324 (1996)
7. Guthaus., M.: MiBench: A free commercially representative embedded benchmark suite. In: 4th Annual IEEE Workshop on Workload Characterization (WWC 2001), Austin, TX, USA, pp. 3–14 (2001)
8. Lee, C., Potkonjak, M., Smith, M.: MediaBench: A tool for evaluating and synthesizing multimedia and communications systems. In: 30th International Symposium on Micro architecture (MICRO-30, 1997), Research Triangle Park, NC, USA, pp. 330–335 (1997)
9. Mahajan, A., Ali, M.S.: Hybrid Evolutionary Algorithm based solution for Register Allocation for Embedded Systems. J. of Computer Programming 3(6), 59–65 (2008)
10. Mahajan, A., Ali, M.S.: Hybrid Evolutionary Algorithm for Graph Coloring Register Allocation. In: IEEE World Congress on Computational Intelligence, Hong Kong, pp. 1162–1167. IEEE Computer Society, Los Alamitos (2008)
11. Scholz, B., Eckstein, E.: Register allocation for irregular architectures. In: Joint Conference on Languages, Compilers and Tools for Embedded systems & software and Compilers for Embedded systems (LCTES 2002-SCOPES 2002), Berlin, Germany, pp. 139–148. ACM Press, New York (2002)
12. Smith, M., Holloway, G.: Graph-coloring register allocation for irregular architectures, Technical report, Harvard University (2000)
13. Topcuoglu, H., Demiroz, B., Kandemir, M.: Solving the register allocation problem for embedded systems using a hybrid evolutionary algorithm. IEEE Transaction on Evolutionary Computation 11(5), 234–620 (2007)
14. Machsuif website,
    http://www.eecs.harvard.edu/hube/research/machsuif.html
15. SNU real-time benchmarks suit,
    http://archi.snu.ac.kr/realtime/benchmark/

Extended Pawlak's Flow Graphs and Information Theory

Huawen Liu[1], Jigui Sun[1], Huijie Zhang[3], and Lei Liu[1,2,*]

[1] College of Computer Science and Technique, Jilin University
Changchun 130012, P.R. China
Huaw.Liu@gmail.com, JgSun@jlu.edu.cn, Liulei@jlu.edu.cn
[2] Key Laboratory of Symbolic Computation and Knowledge Engineering
of Ministry of Education, Changchun 130012, P.R. China
[3] Department of Computer Science, Northeast Normal University
Changchun 130021, P.R. China
Zhanghj167@nenu.edu.cn

Abstract. Flow graph is an effective and graphical tool of knowledge representation and analysis. It explores dependent relation between knowledge in the form of information flow quantity. However, the quantity of flow can not exactly represent the functional dependency between knowledge. In this paper, we firstly present an extended flow graph using concrete information flow, and then give its interpretation under the framework of information theory. Subsequently, an extended flow graph generation algorithm based on the significance of attribute is proposed in virtue of mutual information. In addition, for the purpose of avoiding over-fitting and reducing store space, a reduction method about this extension using information metric has also been developed.

Keywords: Flow graph, knowledge representation, decision table, information theory, decision analysis.

1 Introduction

Knowledge is the symbolic representation of aspects of objects in the world. It stands for our epistemic degree about the world. How to represent knowledge and reason with it is at the heart of most AI systems. As Levesque and Lakemeyer defined in their book [15], knowledge representation is the field of study within AI concerned with using formal symbols to represent a collection of propositions believed by some putative agents. Over the past decades, a growing number of knowledge representation technologies and methodologies have been addressed for a wide variety of applications [10]. For example, logic, tabular, rule and frame are classical and familiar representation forms, while semantic nets and ontologies are newly emerging knowledge representation technologies. The specific form of knowledge representation depends on various factors about problems on hand. For different problems, an appropriate knowledge representation method should

---

* Corresponding author.

M.L. Gavrilova et al. (Eds.): Trans. on Comput. Sci. V, LNCS 5540, pp. 220–236, 2009.
© Springer-Verlag Berlin Heidelberg 2009

be selected, so that experts can easily elicit, describe and comprehend knowledge. Since graphical representation has intuitional and easily understood characteristics, it is very popular in practice. Along with Bayesian network, flow graph is also a graphical one and component for representing knowledge base.

Flow graph (for a short, FG) introduced by Pawlak in his recent paper [21] is a new mathematical model to represent and analyze knowledge in databases. Pawlak's flow graph depicts dependent relation among data in virtue of information flow distribution. Given a FG, the dependent relation between nodes, i.e., knowledge, can be accurately calculated by the flow capacity passed through them. Due to its close relationship with probability theory, FG has been investigated with several theories (e.g., rough sets, decision systems, Bayes' theorem and granular computing). Moreover, the information flow distributions in FG are accord with Bayes' formula and abide by flow conservation equations [22]. More details about FG can be referred to related literatures (see, e.g., [22,23,24,25,26,27,28,29,30,31]). These works, however, pave the way for its application in various fields. For example, Pawlak applied FG into voting analysis [29] and conflict analysis [30], while Kostek and Czyzewski [14] successfully utilized FG model to improve retrieval efficiency in musical metadata retrieval.

Unlike flow networks [11] and control flow graphs [1] that study optimal flow, FG is a quantitative network, and it works under the context of Bayes' theorem. Although it has various advantages, several deficiencies are also embedded in FG. For example, the computing of the dependent relation between two nodes in pair-wise way is really intractable, because the number of nodes is often large. Additionally, if there are two flows passing through nodes a and b and between b and c, respectively, then nodes a and c are necessarily dependent with each other according to Bayes' theorem. However, this is not always true in practice, and nothing may exist between a and c. This implies that the flow distributions can not exactly represent the dependent relation among nodes. To tackle with these troublesome issues, Sun et al. [33] proposed an extension of flow graph (EFG). It rests on the fact that qualitative analysis, as well as quantitative one, also plays an important role in knowledge representation and data analysis, for they can bring more reasonable outcomes to data analysis.

Different from our previous work, where the relationship between Granular computing (GrC) and EFG have been discussed [17], this paper mainly focuses on some issues about EFG, such as reduction and generation of EFG. In real-world, databases are usually with large size. It is impossible to represent them using a great number of nodes in EFG, which will results in both intricate structure and high computational cost. Moreover, the complex representation will be hard to be understood by users and departure from the original idea of knowledge representation. Hence, reduction operation must be carried out over EFG. In this paper, we will cope with this problem using information theory. For the structure of EFG, the arranged order of layers will make a great effect on its reduction. Different tactic orders will yield different structures, which directly determine the inference efficiency and performance of EFG in the future. To obtain an optimal arranged order, we will also employ mutual information to measure the significance of attributes. Based on this measurement, a generation algorithm about EFG is proposed.

The structure of the rest of this paper is organized as follows. Section 2 briefly reviews related work about the state of the art of FG. Section 3 presents necessary notions of decision system and Pawlak flow graphs. In Section 4, an extension of flow graphs based on objects and its interpretations are given. Section 5 mainly offers the relationship between EFG and information theory. Under the context of information metric, the significance of attributes and generation method of EFG are involved in Section 6. Section 7 discusses the issue of reduction in the interest of information-based measurement. Finally, some concluding remarks are showed at the end of this paper.

2   Related Work

Since flow graph has been introduced by Pawlak, many scholars have made an attempt on it and various promising works have been down in the past years. In this section, we will briefly review the state of the art of Pawlak flow graphs.

In [4], Butz et al. demonstrated that FG is a special case of Bayesian network and pointed out that the time complexity of the traditional inference algorithm in FG is exponential. Thus, they introduced an efficient inference method which eliminates variables (i.e., layers) one by one, not all variables at a time. As a result, the reasoning process will be ended in polynomial time [5]. Moreover, a variable elimination order, which is measured by the number of input and output for each node of variable, was investigated in [3].

After introduced the concept of EFG [33], Sun et al. later integrated EFG with granular computing together, for they have common characteristics in structure aspect [17]. Ascribing to composition and decomposition operations between granules, the inference and reformation processes in EFG can be implemented freely and easily under the partition model. Meanwhile, a reduction of EFG can be obtained by using a hierarchy way (i.e., from top to bottom). However, the significance of layer has not been concerned, which will directly determined the performance of the obtained reduction.

Due to its crisp property, FG can not directly be utilized to represent fuzzy information systems. Hence, Mieszkowicz-Rolka and Rolka [19] generalized FG into a fuzzy one, called fuzzy flow graph, to suit for representing and analyzing decision tables with fuzzy attributes. In this kind of FG, each node in layer refers to one linguistic value of the corresponding attribute. In order to satisfy the flow conservation equations, a T-norm operator was chosen in calculating the information flow between nodes. Contrastively, the certainty and strength of path were also extended.

As mentioned above, rule is also a kind of knowledge representation and owing to its easy comprehension and interpretation, it has been widely used in practice. Since FG excels at graphical representation, many endeavors have been attempted to build the relationship between rules and FGs. For example, Pattaraintakorn et al. [20] generalized rules for ordinal prediction, and then redefined the graphical representation of rules with ordinal prediction while considering the measures in FG. Recently, Chitchareon and Pattaraintakorn [7] revealed the relationship between association rules and flow graphs. In this method, a path $[xyz]$ in FG is an association rule $xy \rightarrow z$ whose support and confidence are $\varphi(xyz)$ and one, respectively, if its certainty is completely determinate, that is, $cer(xyz) = 1$.

Chan and Tsumoto [6] revealed the relationship between decision rules and FGs using multiset decision tables. In multiset decision table, each row corresponds to one path in FG. Under this scheme, the multiset decision table is a minimal representation of FG. The advantage lies in the fact that it facilitates the learning of decision rules from inconsistent data, and the obtained rule set is a minimal one.

For the application of FG, Kostek and Czyzewski [14,9] successfully applied FG to analyze the dependent relation between meta-data in musical databases. Their purpose is to enable effective and efficient access to such information and release the limitation of online resources. In temporal data mining, one of important aspects is the analysis of data changing in time. For succeeding data, its trend is determined by the anterior models, and this is linked inside by means of functional dependency. To obtain this functional relation to predict future behavior of modeled systems, Suraj and Pancerz [34] exploited FG to mine prediction rules to describe the changes of components in the consecutive time windows of a temporal information system.

3   Basic Concepts

In this section, some basic concepts about decision table and flow graph will be recalled briefly. More relevant notations can be consulted relative literatures (see, e.g., [26,29]).

3.1   Decision Tables

Formally, An information system is $S = (U, A)$, where U and A are non-empty finite sets called *universe* and *attributes*, respectively. In a slight abuse of notations, for any attribute $a \in A$, it is denoted as a function $a : U \rightarrow V_a$, where V_a is the *value domain* of attribute a (i.e., the set of all values). In information system, an important concept called *indiscernibility relation* is defined as a binary relation $I(B)$ on U such that $(x, y) \in I(B)$ if and only if $a(x) = a(y)$ for every $a \in B$, where $B \subseteq A, a \in B, x, y \in U$ and $a(x)$ denotes the value of attribute a for element x.

Decision table is a special kind of information system. It is denoted as $DT = (U, C, D)$, where U, C and D are *universe, condition attributes* and *decision attributes*, respectively. Usually, $C \cup D = A$ and $C \cap D = \emptyset$. Let $DT = (U, C, D)$ be a decision table, each row in DT determines a decision rule $C \rightarrow D$, which means that specific decisions should be taken when conditions are satisfied. The numerical values $supp(C, D) = |C(x) \cap D(x)|$ and $str(C, D) = supp(C, D)/|U|$

are respectively called *support* and *strength* of $C \to D$, where $|X|$ denotes the cardinality of set X. For each decision rule, there are two major factors associated, i.e., *certainty* and *coverage*. The certainty and coverage of $C \to D$ are defined as $cer(C, D) = supp(C, D)/|C(x)|$ and $cov(C, D) = supp(C, D)/|D(x)|$, respectively.

In decision table, decision rules with the same conditions but different decisions are called *inconsistent*; otherwise they are *consistent*. Decision table is inconsistent if it contains inconsistent decision rules; otherwise the table is consistent. A rule base consists of all decision rules induced from a decision table. It is also called decision algorithm from the view of logic, where the basic element is *descriptor* (a, v) (namely, attribute-value pair) and formulas are built up from these descriptors by means of logical connectives in the standard way. Given a descriptor (a, v), its *meaning* are those elements $x \in U$ such that $a(x) = v$. The interested reader is advised to consult the references (e.g., [26,13]) to obtain more details about decision algorithm, for it is out of the scope of this paper.

For the sake of simplicity, we assume that decision table is always consistent and has a single decision attribute, i.e. $D = \{d\}$.

3.2   Flow Graphs

Flow graph (FG) is a directed, acyclic, finite graph $G = (N, B, \varphi)$, where N is a set of *nodes*, $B \subseteq N \times N$ is a set of *directed branches*, $\varphi : B \to R^+$ is a flow function and R^+ is non-negative reals [26].

In FG, if $(n, n') \in B$, node n is an *input* of n' and n' is an output of n reversely. In addition, $\varphi(n, n')$ is the *throughflow* passing through the branch $(n, n') \in B$ from n to n'. The sets $I(n) = \{n'|(n', n) \in B\}$ and $O(n) = \{n'|(n, n') \in B\}$ denote as inputs and outputs of node n, respectively. For graph G, its inputs and outputs are $I(G)=\{n \in N|I(n) = \emptyset\}$ and $O(G)=\{n \in N|O(n) = \emptyset\}$. The *inflow* and *outflow* of node n are represented as

$$\varphi_+(n) = \sum_{n' \in I(n)} \varphi(n', n) \quad \text{and} \quad \varphi_-(n) = \sum_{n' \in O(n)} \varphi(n, n').$$

The throughflow of FG G is $\varphi(G) = \sum_{n \in I(G)} \varphi_-(n) = \sum_{n \in O(G)} \phi_+(n)$.

A *normalized flow graph* is a directed, acyclic, finite graph $G = (N, B, \sigma)$, where N and $B \subseteq N \times N$ defined as above, $\sigma : B \to (0, 1)$ is a normalized flow function of branches. The *strength* of $(n, n') \in B$ is $\sigma(n, n') = \varphi(n, n')/\varphi(G)$. Similarly, the *inflow* and *outflow* of node n in G are

$$\sigma_+(n) = \varphi_+(n)/\varphi(G) \quad \text{and} \quad \sigma_-(n) = \varphi_-(n)/\varphi(G),$$

respectively. The *certainty* and *coverage* of branch (n, n') are defined as $cer(n, n') = \sigma(n, n')/\sigma(n)$ and $cov(n, n') = \sigma(n, n')/\sigma(n')$, where $\sigma(n) \neq 0$ and $\sigma(n') \neq 0$.

Example 1. Consider a decision table $DT = (U, C, D)$ given as Table 1, where $U = \{1, .., 11\}$, $C = \{c1, .., c4\}$ and $D = \{d\}$. As a matter of fact, this decision

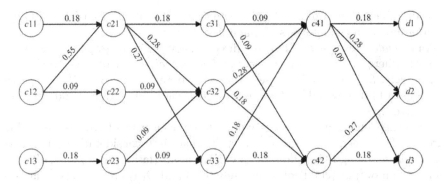

Fig. 1. A normalized flow graph $G = (N, B, \phi)$ of the decision table DT

table is induced from the *Iris Plants* (Iris) database, which is well-known in machine learning community and freely accessible at the UCI machine learning repository [2]. After numerical attributes in the original Iris have been discretized and the same rows have been removed, the data in the Iris is shown as Table 1.

Table 1. The decision table $DT = (U, C, D)$

Id	c1	c2	c3	c4	d	Id	c1	c2	c3	c4	d	Id	c1	c2	c3	c4	d
1	1	1	2	1	1	5	2	1	2	1	2	9	2	2	2	1	3
2	1	1	3	1	1	6	2	1	2	2	2	10	3	3	2	2	3
3	2	1	1	1	2	7	2	1	3	1	2	11	3	3	3	2	3
4	2	1	1	2	2	8	2	1	3	2	2						

Fig. 1 is the normalized flow graph $G = (N, B, \sigma)$ of DT, where $N = \{c11, ..., d3\}, B = \{(c11, c21), ..., (c42, d3)\}$. Node $c11$ denotes the descriptor $(c1, 1)$, i.e., the value of attribute $c1$ is one.

For node $c21 \in N$, its input and output are $\{c11, c12\}$ and $\{c31, c32, c33\}$. The throughflow of branch $(c11, c21) \in B$ is 0.18. In terms of definition, $\sigma(c21) = 0.18 + 0.55 = 0.18 + 0.28 + 0.27 = 0.73$. The strength, certainty and coverage of $(c21, c31)$ are $\sigma(c21, c31) = 0.18, cer(c21, c31) = 0.18/0.73 = 0.25$ and $cov(c21, c31) = 0.18/0.18 = 1$. Node sequence $c11, c21, c33, c41, d1$ is a decision rule $c11, c21, c33, c41 \to d1$. □

4   Extended Flow Graphs

As mentioned above, FG is a quantification graph, that is, it represents simply the relations among nodes by virtue of flow distribution of information. However, it is not sufficient to depict concretely and exactly the relations among nodes in quantitative way. As an example, a decision rule $c11, c21, c33, c42 \to d2$ can be induced from the FG in Fig. 1. Actually, this rule can not be derived from the decision table DT in Table 1. To alleviate this problem, an extension of FG has been proposed in the interests of qualitative factors.

An *extended flow graph* (EFG) is a directed, acyclic, finite graph $G = (E, N, B, \phi, \alpha, \beta)$, where E and N are the sets of *flow capacities* (or objects) and nodes respectively. $B \subseteq N \times N$ is the set of directed branches, $\phi : B \to 2^E$ is *capacity function* of branches and $\alpha, \beta \to [0, 1]$ are thresholds of certainty and decision, respectively.

From this definition, the capacity of each branch is a set of objects which flow through the branch. This means that EFG takes into consideration qualitative factors. Resembling FG, node n in EFG is an input (father) of n', if $(n, n') \in B$, and n' is an output (child) of n reversely. $I(n)$ and $O(n)$ are the sets of fathers and children of node n respectively. In addition, node n is called the *root* if $I(n) = \emptyset$, while n is a *leaf* if $O(n) = \emptyset$. n is an *condition node* if it is neither the root nor a leaf.

Given an EFG G, the input and output capacities of node n are defined as

$$\phi_+(n) = \bigcup_{n' \in I(n)} \phi(n', n) \quad \text{and} \quad \phi_-(n) = \bigcup_{n' \in O(n)} \phi(n, n'), \qquad (1)$$

respectively. If n is an condition node, then we have $\phi(n) = \phi_+(n) = \phi_-(n)$. In addition, $\phi(n, n') = \phi_-(n) \cap \phi_+(n') = \phi(n) \cap \phi(n')$ also holds on. This implies a fact that if an object flows through branch (n, n'), it must also pass through both n and n', and vice versa. The *strength* of branch (n, n') is referred to $|\phi(n, n')|$. Moreover, its *certainty* and *coverage* are denoted as

$$cer(n, n') = |\phi(n, n')| / |\phi(n)| \quad \text{and} \quad cov(n, n') = |\phi(n, n')| / |\phi(n')|, \qquad (2)$$

where $|X|$ is the cardinality of X, $\phi(n) \neq \emptyset$ and $\phi(n') \neq \emptyset$.

A *directed path* from n to n', denoted by $[n...n']$, is a directed sequence of nodes $n_1, ..., n_m$, where $(n_i, n_{i+1}) \in B$ for $1 \leq i \leq m - 1$, $n_1 = n, n_m = n'$ and $\bigcap \phi(n_i, n_{i+1}) \neq \emptyset$. In a similar vein, the *support, certainty* and *coverage* of the path $[n_1...n_m]$ are represented as

$$\begin{aligned} sup(n_1...n_m) &= \bigcap \phi(n_i, n_{i+1}), \\ cer(n_1...n_m) &= |\phi(n_1...n_m)| / |\phi(n_1...n_{m-1})| \\ cov(n_1...n_m) &= |\phi(n_1...n_m)| / |\phi(n_m)|, \end{aligned} \qquad (3)$$

respectively, where $\phi(n_1...n_{m-1}) \neq \emptyset$ and $\phi(n_m) \neq \emptyset$.

This extension of FG can bring several advantages for data analysis. First of all, EFG has more powerful and robust capability of representation and analysis than FG. If we only concern the quantity of flow capacity, rather than specific objects flowing through branches, then an EFG $G = (E, N, B, \phi, \alpha, \beta)$ can be transformed into corresponding FG $G = (N, B, \varphi)$. That is to say, $\phi(n, n') = |\varphi(n, n')| / |E|, \alpha = 0$ and $\beta = 0$. Additionally, an approximate FG can be obtained by adjusting the parameters α and β in EFG.

Secondly, EFG has the same representation power with corresponding decision table [33]. Given an EFG $G = (E, N, B, \phi, \alpha, \beta)$, it can be easily transformed

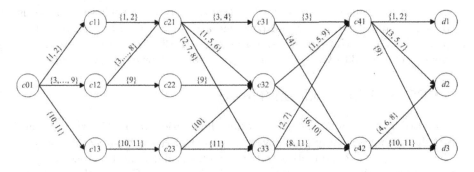

Fig. 2. The EFG $G = (E, N, B, \phi, \alpha, \beta)$ induced from the decision table DT in Table 1

into a decision table $DT = (U, C, D)$, and vice versa. To be more specific, for the EFG G, its total capacity of G is the universe in DT, i.e., $E = U$. Each node $n \in N$ means an attribute-value pair (a, v) in DT, and its capacity of flow is equal to the meaning of the pair, namely, $\phi(n) = \{x \in U | a(x) = v\}$, where $a \in C \cup D$ and $v \in V_a$. Additionally, branch $(n, n') \in B$ denotes that its corresponding attribute-value pairs associated with each other.

Granular computing (GrC) is a practical methodology of problem solving deeply rooted in human mind [38]. It decomposes a hard and complex problem into several easy ones to solve them individually and then combines the solutions into the whole, by effectively using levels of granularity. EFG, however, is also intrinsically consistent with GrC, for they shares many common features in structural facet. Contrastively, each node $n \in N$ and its flow capacity $\phi(n)$ in EFG represents a element granule $(n, \phi(n))$, and the composition and decomposition operation can be implemented by the branches in EFG. More details about the relationship between EFG and GrC can be referred to literatures (e.g., [17]).

EFG is similar to the notation of information map proposed by Skowron and Synak [32]. Both of them represent information or knowledge by means of states (or nodes) and associated relations among them. However, they distinguish with each other at two aspects: binary relation and information function.

For the convenience of discussion, hereafter we assume that nodes in an EFG with same attribute are arranged in same layer and the whole EFG consists of several layers. The first layer is the root r, while the last one is the leaves. In the same layer, there does not exist any branch among nodes and each object only flows through one node. The flow capacity of any branch is not empty.

Example 2. (cont.) Given the decision table $DT = (U, C, D)$, shown as Table 1, its corresponding EFG $G = (E, N, B, \phi, \alpha, \beta)$ is depicted as Fig. 2. The total flow capacity of G is $E = \{1, ..., 11\}$. In this EFG, each layer is corresponding to an attribute in DT, except $c01$. Contrastively, the decision attribute is denoted as the leaf layer. The root and leaf nodes in the EFG G are $N_r = c01$ and $N_d = \{d_1, d_2, d_3\}$ respectively.

For branch $(c12, c21)$, its capacity is $\phi(c12, c21) = \{3, ..., 8\}$, and the flow capacity of $c21$ is $\phi(c21) = \{1, 2\} \cup \{3, ..., 8\} = \{3, 4\} \cup \{1, 5, 6\} \cup \{2, 7, 8\} =$

$\{1, ..., 8\}$. Node sequence $p = c11, c21, c33, c41, d1$ is a path whose capacity is $\phi(p) = \{2\}$, and its certainty and coverage are $cer(p) = 1$ and $cov(p) = 1/2$, respectively. This path is the decision rule $c11, c21, c33, c41 \rightarrow d1$. However, the sequence $c11, c21, c33, c42 \rightarrow d2$ is not a path, for its flow capacity is empty.   □

5   EFG and Information Theory

Information plays a vital role in connecting the physical world with the abstract world. It is defined as any property or attribute of the natural world that can be generally abstracted, quantitatively represented, and mentally processed [36]. As one of knowledge representation and analysis tools, EFG is not the exception and it can also be measured by information metric. In this subsection, we will discuss the relationship between EFG and information theory.

5.1   Information Theory

Since information theory is capable of quantifying the uncertainty of random variables and measures the amount of information shared by them effectively, it has been widely used in many fields, such as science, engineering, business [8]. In spite of having various definitions and meanings, the content of information, however, can be measured by the unified scheme: probability theory [37].

Information entropy is perhaps the most fundamental concept in information theory. For a random discrete variable X, its *entropy* $H(X)$ is defined as

$$H(X) = - \sum_{x \in X} p(x) \log p(x) \tag{4}$$

where $p(x)$ is probability density function of X. Given the observing values of Y, the *conditional entropy* $H(X|Y)$ of X with respect to Y is

$$H(X|Y) = - \sum_{y \in Y} \sum_{x \in X} p(x, y) \log p(x|y) \tag{5}$$

where $p(x|y)$ is the posterior probabilities of X given the values of Y, The conditional entropy $H(X|Y)$ quantifies the remaining uncertainty of X when Y is known. If X is completely dependent on Y, then $H(X|Y) = 0$. This means that no more other information is required to describe X when Y is known. On the contrary, $H(X|Y) = H(X)$ holds if they are independent. That is to say, knowing Y will do nothing to observe X.

To illustrate the common information between variables, a concept of *mutual information* is addressed. For two random discrete variables X and Y, their mutual information $I(X; Y)$ is

$$I(X, Y) = H(X) - H(X|Y) \tag{6}$$

From equation (6), we have the fact that if X and Y are closely related, $I(X; Y)$ is very high. Otherwise, $I(X; Y) = 0$ implies that these two variables are totally unrelated or independent with each other.

Analogically, if Z has been given, the *conditional mutual information* of X and Y, denoted as $I(X;Y|Z) = H(X|Z) - H(X|Y,Z)$, represents the quantity of information shared between X and Y when Z is known. It can be seen that the value of $I(X;Y|Z)$ implies Y brings information about X which is not already contained in Z. In other words, the larger value is, the more information has.

5.2 Information of EFG

In an EFG $G = (E, N, B, \phi, \alpha, \beta)$, the basic element is node or branch. For each node $n \in N$, its flow capacity implies the representation ability. Thus, the information embodied in node can also be measured by its flow capacity $\phi(n)$. Specifically, let $p(n) = |\phi(n)|/|E|$ be the probability of flow capacity of n, the information amount of the node n is encoded as

$$I(n) = -p(n)log(p(n)).$$

Furthermore, the information encompassed by the EFG G is denoted as

$$I(G) = \sum_{n \in N} I(n).$$

This measurement, however, is another form of information entropy, and competent for quantifying the uncertainty of G. If the distributions of flow capacity of nodes are complete disorder or randomization, the information value $I(G)$ of G is maximal. While the maximum one occurs if the probability distributions are equal. Moreover, one can obtain the following property in a straightforward way.

Property 1. *Let $S = (U, A)$ be an information system, $G = (E, N, B, \phi, \alpha, \beta)$ and $G' = (E, N', B', \phi', \alpha, \beta)$ are two different EFGs of the information system S, then we have $I(G) = I(G')$.*

Although $I(G)$ is capable of measuring the information amount embodied in G, it does not consider the interaction and dependence between nodes. Thus, it is inappropriate to the decision information in EFG. As mentioned above, mutual information is competent for quantifying the dependent relation between variables. In EFG, the dependent relation between nodes is represented as branch and its strength is determined by the corresponding flow capacity. Thus, we can exploit mutual information to measure the dependent relations between condition nodes and decision nodes.

For the sake of discussion, the nodes in N are grouped into three disjointed categories, i.e., the *root node* N_r, *internal nodes* N_i and *decision nodes* N_d, where $N_r \cup N_i \cup N_d = N$. Thus, EFG is also represented as $G = (E, N_c \cup N_d, B, \phi, \alpha, \beta)$, where N_c is the set of *condition nodes* (i.e., $N_c = N_r \cup N_i$). For each condition node $n \in N_c$, its mutual information $I(n; N_d)$ with respect to decision nodes N_d is defined as

$$I(n; N_d) = I(n) - I(n|N_d).$$

If the value $I(n; N_d)$ is very high, the impact of knowledge n performing on N_d would be significant, as users are making a decision on the basis of available knowledge. For all condition nodes N_c in G, its total dependent relation with decision nodes N_d is $I(N_c; N_d)$, and this is the decision information embodied in G. Similarly, the following property can be obtained easily.

Property 2. *Let* $DT = (U, C, D)$ *be a decision table,* $G = (E, N_c \cup N_d, B, \phi, \alpha, \beta)$ *and* $G' = (E, N'_c \cup N'_d, B', \phi', \alpha, \beta)$ *are two different EFGs of the decision table* DT, *then we have* $I(N_c; N_d) = I(N'_c; N_d)$.

6   Generation of EFG

For the same decision table $DT = (U, C, D)$, there have various forms of EFG in the light of different arranged order of layers (i.e., attributes). Although these different EFGs have same decision information, they will exert great influence to the reduction operation. As a result, different reductions will be produced from these EFGs. To achieve an optimal reduction of EFG, the order of attributes in decision table should be determined on the ground of their significant degrees in making decisions, before its corresponding EFG has been generated.

As a matter of fact, the issue of the importance order about attributes, which is also called feature ranking or selection [12], is a crucial problem in data mining. For feature selection, interesting readers can consult relative literatures (e.g., [18,12]) to get more information. Despite that there are various outstanding feature selection algorithms, no one method predominates over others in all situations. This tells us a fact that different methods should be adopted to suit for specific problems at hand. For example, Lin and Yin [16] pointed out that the attribute with the most values is more important than others. While Wang in [35] argued that the measurements of the maximal certainty and coverage degrees also play vital roles in classification, and the attribute with maximal certainty or coverage has higher superiority.

In decision analysis, decision making mainly rests on the reasoning procedure over available or acquired knowledge (i.e., values of condition attributes). This, however, can be denoted by dependent relations between condition and decision attributes. As mentioned above, mutual information is an effective metric to quantify the dependency and uncertainty between attributes. Moreover, the larger the mutual information $I(c, d)$, the higher the dependent degree between c and d, where $c \in C$ and $d \in D$ are condition and decision attributes respectively. Based on this principle, the generation algorithm of EFG is shown as Alg. 1.

This generation algorithm works in a straightforward way. At first, it estimates mutual information for each candidate attribute in C and sorts them in descending order. During each repetition stage, the attribute $c \in C$ with the highest priority is chosen. Then, new nodes and branches, as well as flow capacities induced from c, are inserted into N and B. After that, c will be removed from the candidate list C and added into the selected attribute list SA. This procedure will be terminated if there has no available candidate attributes or the information of EFG will not be changed. At the end of this algorithm, the relationships between the root and the first selected attribute and between the decision nodes and the last selected attribute will be updated.

Algorithm 1. The EFG generation Algorithm

Input : A decision table $DT = (U, C, \{d\})$;
Output: An EFG $G = (E, N, B, \phi, \alpha, \beta)$ induced from DT;
Initialize relative parameters: $E = U, N = N_r \cup N_d, \alpha = \emptyset, \beta = \emptyset$, where
$N_d = \{d_i\}, d_i \in V_d$, is leaves and $N_r = \{r\}$ is the root node ;
for *each condition attribute* $c \in C$ **do**
  | Calculate mutual information $I(c; d)$ and sort them in descending order ;
end
$SA = \emptyset$; // SA is the set of selected attributes ;
while $C \neq \emptyset$ **do**
  | Select the attribute $c \in C$ with the highest $I(c; d)$;
  | **for** *each value* $v \in V_c$ **do**
  | | $N = N \cup \{v\}; \phi(v) = \{x | c(x) = v\}$;
  | | $B = B \cup \{(v', v)\}; \phi(v', v) = \{x | c(x) = v, c'(x) = v'\}$, where c' is the last
  | | selected attribute in SA and $n \in V_c'$;
  | | $C = C - \{c\}; SA = SA \cup \{c\}$;
  | **end**
end
$B = B \cup \{(r, v)\}; \phi(r, v) = \phi(v)$, where v is the first selected attribute in SA;
$B = B \cup \{(v, d_i)\}; \phi(v, d_i) = \{x | c(x) = v, d(x) = d_i\}$, where v is the last selected
attribute in SA;

The computational complex of this algorithm is $O(mk^2)$, where m is the number of attributes and k is the maximal number of values of attributes. In practice, not all attributes are important to decision analysis, and some attributes have no any contribution, except to distract users. However, Alg. 1 does not take into consideration interaction between attributes. This problem can be circumvented by imposing constraints on it. For example, the mutual information in the first *for* statement can be replaced with condition mutual information $I(c; d | SA)$. Additionally, $I(SA; d) = I(C; d)$ becomes one of components of stopping condition. After minor modifications, this algorithm can also be utilized to implement the function of self-regulation to adapt to dynamic situations.

Example 3. (cont.) For the decision table $DT = (U, C, D)$ in Table 1, after being calculated mutual information, their values of $c1, c2, c3$ and $c4$ are $I(c1; d) = 1.06$, $I(c2; d) = 0.85, I(c3; d) = 0.20$ and $I(c4; d) = 0.20$, respectively. Thus, the selection order is $c1, c2, c3, c4$, and its EFG G (Fig. 1) is generated at end. □

7 Reduction of EFG

Generally, decision table in real-world is very large and filled with thousands of data represented by a great number of attributes. This will raise a problem that the cost of inference in its corresponding EFG is expensive and the storage capacity is high. Additionally, the induced EFG is too complex to be understood and interpreted by users. Therefore, it is necessary to simplify EFG without losing its decision capability, after it has be yielded from the given decision table. Actually, it is more preferable to perform inference on a reduction of EFG when decisions are making. Meanwhile, the over-fitting situations in the reduction will be reduced. Before we delve into the details of reduction algorithms, let us firstly turn our concerns on the concept of reduction.

When an EFG is derived from a decision table, it also embodies decision information. These information, however, should be preserved in its reduction, otherwise the decision information would be lost.

Definition 1. *Given two EFGs $G = (E, N_c \cup N_d, B, \phi, \alpha, \beta)$ and $G' = (E', N'_c \cup N'_d, B', \phi, \alpha, \beta)$ derived from the same decision table, where N_c and N_d represent conditional nodes and leaves, respectively. G' is called a reduction of G, if the decision information embodied in G' is the same with those in G and $N'_c \subseteq N_c$.*

This definition implies that the reduction must have the same decision information with the original EFG, while the number of its nodes is less than those of the original one. For decision information, there are various measurements, e.g., consistent factor [26], dependency and information entropy [8]. We have discussed the reduction under the context of consistent measurement in [33].

From the view of information theory, the information element in EFG is node and branch. Thus, for a given EFG G and its reduction G', $I(N_c; N_d) = I(N'_c; N_d)$ holds on the ground of Def. 1, where N'_c is the condition nodes in G' and $N'_c \subseteq N_c$. As a result, the reduction problem is immigrated into the one to achieve a minimal subset of condition nodes whose mutual information is equal to those of the original EFG with respect to decision nodes.

In order to acquire a minimal subset, two strategies are available. One is sequential forward selection, and another is sequential backward elimination [18]. The former begins with an empty set, and adds an condition node into the selected set at a time, where the selected node has higher mutual information than others. However, the latter works in a contrary way. It starts at the entire condition nodes N_c, and nodes which have no any contribution to $I(N_c; N_d)$ will be removed one by one, until the mutual information is changed. Although there have respective characteristics, we here adopt the sequence forward selection strategy in our reduction algorithm.

In selecting individual condition node, the node $n \in N_c$ with the largest $I(n; N_d)$ is preferable, for it contains more decision information than others. However, this is not always true in practice. The reason is that the information contained in n may be already encompassed by the selected nodes N'_c. To alleviate this problem, conditional mutual information $I(n; N_d|N'_c)$ should be taken as the selection criterion, and those nodes with the largest $I(n; N_d|N'_c)$ will be culled firstly. More specifically, the details of reduction method using information metric are presented as Alg. 2.

In this algorithm, the most cost computation is the *for* statement, and it will take $O(m)$ to estimate conditional mutual information $I(n; N_d|N'_c)$ for every node, where $m = |E|$. Let k be the total number of nodes in G, the *while* statement will run at most k times before it stops. Thus, the total computational cost of this algorithm is $O(mk^2)$.

Example 4. (cont.) For the EFG G shown as Fig. 2, its embodying decision information is $I(N_c; N_d) = 1.44$. In the first iteration, the node $c11$ has the most

Algorithm 2. EFG reduction algorithm using information metric

---

Input : An EFG $G = (E, N_c \cup N_d, B, \phi, \alpha, \beta)$;
Output: A reduction $G' = (E, N_c' \cup N_d, B', \phi, \alpha, \beta)$ of G;
Initialize relative parameters: $N_c' = N_r$; $B' = \emptyset$, where N_r is the root node ;
Calculate the decision information $I(N_c; N_d)$ embodied in G ;
while $I(N_c'; N_d) \neq I(N_c; N_d)$ **do**
  for *each condition node n in N_c* **do**
  | Calculate its condition mutual information $I(n; N_d|N_c')$;
  end
  Select the node $n \in N_c$ with largest $I(n; N_d|N_c')$ and
  $N_c' = N_c' \cup \{n\}$; $N_c = N_c - \{n\}$;
end
Update the set of branches B' in terms of N_c' and the order of layers in G;

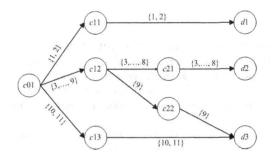

Fig. 3. A reduction G' of G (Fig. 2) using Alg. 2

information whose value is $I(c11; N_d) = 0.45$. After $c11$ has been selected, $c13$ or $c23$ will be picked out in the next iteration, for their condition mutual information $I(c13; N_d|c11) = I(c23; N_d|c11) = 0.34$ are the highest. If $c13$ is chosen, $I(c23; N_d|c11, c13) = 0$. In a similar vein, nodes $c12, c21$ and $c22$ will be chosen one after the other. At this time, the information value of N_c' is $I(N_c'; N_d) = 1.44$. Thus, the reduction algorithm is terminated and the final reduction EFG G' is illustrated as Fig. 3. □

8 Conclusions

In this paper, we firstly discussed a graphical framework of knowledge representation: extended Pawlak's flow graph (EFG). The main predominance of EFG lies in that it can represent and analyze knowledge in both quantitative and qualitative manners. Then the relationship between EFG and information theory is presented. Under the context of information metric, the information amount of each node in EFG is denoted by its flow capacity and the dependent degree between condition nodes and decision nodes can be measured by the mutual information between them. In addition, an EFG generation algorithm using attribute significance has been given. On the basis of this EFG, a kind of reduction algorithm about EFG has been developed by virtue of information measurement.

One may note that different selection ordering of nodes may arise a problem of uniqueness in generation and reduction operation, as nodes have equivalent mutual information. In our future work, the relationship between different EFGs induced from the same decision table will be investigated. Moreover, the practical applications of EFGs will also be carried out by using large and real-world datasets.

Acknowledgements

The authors are grateful to anonymous referees for their valuable and constructive comments. This work is supported by the Doctor Point Founds of Educational Department (20060183044) and Science Foundation for Young Teachers of Northeast Normal University(20081003).

References

1. Allen, F.E.: Control flow analysis. ACM SIGPLAN Notices 5(7), 1–19 (1970)
2. Blake, C.L., Merz, C.J.: UCI Repository of Machine Learning Databases (1998), http://www.ics.uci.edu/~mlearn/MLRepository.html
3. Butz, C.J., Yan, W.: Current Trends in Rough Set Flow Graphs. In: Hassanien, A.E., et al. (eds.) Rough Computing: Theories, Technologies and Applications, pp. 152–161. Idea Group Inc. (2007)
4. Butz, C.J., Yan, W., Yang, B.: The Computational Complexity of Inference using Rough Set Flow Graphs. In: Ślęzak, D., Wang, G., Szczuka, M.S., Düntsch, I., Yao, Y. (eds.) RSFDGrC 2005. LNCS, vol. 3641, pp. 335–344. Springer, Heidelberg (2005)
5. Butz, C.J., Yan, W., Yang, B.: An Efficient Algorithm for Inference in Rough Set Flow Graphs. Transaction on Rough Sets V, 102–122 (2006)
6. Chan, C.-C., Tsumoto, S.: On Learning Decision Rules From Flow Graphs. In: Proceedings of North American Fuzzy Information Processing Society (NAFIPS 2007), pp. 655–658 (2007)
7. Chitchareon, D., Pattaraintakorn, P.: Knowledge Discovery by Rough Sets mathematical Flow Graphs and Its Extension. In: Proceedings of the IASTED International Conference on Artificial Intelligence and Applications (AIA 2008), pp. 340–345 (2008)
8. Cover, T.M., Thomas, J.A. (eds.): Elements of Information Theory. Wiley, New York (1991)
9. Czyzewski, A., Szczerba, M., Kostek, B.: Musical Metadata Retrieval with Flow Graphs. In: Tsumoto, S., Słowiński, R., Komorowski, J., Grzymała-Busse, J.W. (eds.) RSCTC 2004. LNCS, vol. 3066, pp. 691–698. Springer, Heidelberg (2004)
10. Davis, R., Shrobe, H., Szolovits, P.: What is a Knowledge Representation? AI Magazine 14(1), 17–33 (1993)
11. Ford, L.R., Fulkerson, D.R. (eds.): Flows in networks. Princeton University Press, New Jersey (1962)
12. Jain, A.K., Duin, R.P.W., Mao, J.: Statistical Pattern Recognition: A Review. IEEE Transactions on Pattern Analysis and Machine Intelligence 22(1), 4–37 (2000)
13. Komorowski, J., Pawlak, Z., Polkowski, L., Skowron, A.: Rough sets: a tutorial. In: Pal, S.K., Skowron, A. (eds.) Rough Fuzzy Hybridization: A New Trend in Decision-Making, pp. 3–8. Springer, Singapore (1999)

14. Kostek, B., Czyzewski, A.: Processing of Musical Metadata Employing Pawlak's Flow Graphs. In: Peters, J.F., Skowron, A. (eds.) Transactions on Rough Sets III. LNCS, vol. 3400, pp. 279–298. Springer, Heidelberg (2004)

15. Levesque, H., Lakemeyer, G. (eds.): The Logic of Knowledge Bases. MIT Press, Cambridge (2000)

16. Lin, T.Y., Yin, P.: Heuristically Fast Finding of the Shortest Reducts. In: Tsumoto, S., Słowiński, R., Komorowski, J., Grzymała-Busse, J.W. (eds.) RSCTC 2004. LNCS, vol. 3066, pp. 465–470. Springer, Heidelberg (2004)

17. Liu, H., Sun, J., Zhang, H.: Interpretation of Extended Pawlak Flow Graphs using Granular Computing. In: Peters, J.F., Skowron, A. (eds.) Transactions on Rough Sets VIII. LNCS, vol. 5084, pp. 93–115. Springer, Heidelberg (2008)

18. Liu, H., Yu, L.: Toward Integrating Feature Selection Algorithms for Classification and Clustering. IEEE Transactions on Knowledge and Data Engineering 17(4), 491–502 (2005)

19. Mieszkowicz-Rolka, A., Rolka, L.: Flow Graphs and Decision Tables with Fuzzy Attributes. In: Rutkowski, L., Tadeusiewicz, R., Zadeh, L.A., Żurada, J.M. (eds.) ICAISC 2006. LNCS(LNAI), vol. 4029, pp. 268–277. Springer, Heidelberg (2006)

20. Pattaraintakorn, P., Cercone, N., Naruedomkul, K.: Rule learning: Ordinal prediction based on rough sets and soft-computing. Applied Mathematics Letters 19(12), 1300–1307 (2006)

21. Pawlak, Z.: Decision algorithms, Bayes' Theorem and Flow Graphs. European Journal of Operational Research 13(6), 181–189 (2002)

22. Pawlak, Z.: The Rough Set View on Bayes' Theorem. In: Pal, N.R., Sugeno, M. (eds.) AFSS 2002. LNCS, vol. 2275, pp. 106–120. Springer, Heidelberg (2002)

23. Pawlak, Z.: Probability, truth and flow graphs. In: Proceedings of the Workshop on Rough Sets in Knowledge Discovery and Soft Computing at ETAPS 2003, pp. 1–9 (2003)

24. Pawlak, Z.: Flow graphs and decision algorithms. In: Proceedings of the 9th International Conference on Rough Sets, Fuzzy Sets, Data Mining and Granular Computing. LNCS, vol. 2639, pp. 1–11. Springer, Heidelberg (2003)

25. Pawlak, Z.: Decision Networks. In: Tsumoto, S., Słowiński, R., Komorowski, J., Grzymała-Busse, J.W. (eds.) RSCTC 2004. LNCS, vol. 3066, pp. 1–7. Springer, Heidelberg (2004)

26. Pawlak, Z.: Some Issues on Rough Sets. In: Peters, J.F., Skowron, A., Grzymała-Busse, J.W., Kostek, B.z., Świniarski, R.W., Szczuka, M.S. (eds.) Transactions on Rough Sets I. LNCS, vol. 3100, pp. 1–58. Springer, Heidelberg (2004)

27. Pawlak, Z.: Decisions Rules and Flow Networks. European Journal of Operational Research 154, 184–190 (2004)

28. Pawlak, Z.: Rough Sets and Flow Graphs. In: Ślęzak, D., Wang, G., Szczuka, M.S., Düntsch, I., Yao, Y. (eds.) RSFDGrC 2005. LNCS, vol. 3641, pp. 1–11. Springer, Heidelberg (2005)

29. Pawlak, Z.: Flow Graphs and Data Mining. In: Peters, J.F., Skowron, A. (eds.) Transactions on Rough Sets III. LNCS, vol. 3400, pp. 1–36. Springer, Heidelberg (2005)

30. Pawlak, Z.: Some remarks on conflict analysis. European Journal of Operational Research 166(3), 649–654 (2005)

31. Pawlak, Z.: Decision Trees and Flow Graphs. In: Greco, S., Hata, Y., Hirano, S., Inuiguchi, M., Miyamoto, S., Nguyen, H.S., Słowiński, R. (eds.) RSCTC 2006. LNCS, vol. 4259, pp. 1–11. Springer, Heidelberg (2006)

32. Skowron, A., Synak, P.: Reasoning Based on Information Changes in Information Maps. In: Proceedings of the 9th International Conference on Rough Sets, Fuzzy Sets, Data Mining and Granular Computing (RSFSGrC 2003). LNCS, vol. 2639, pp. 229–236. Springer, Heidelberg (2003)
33. Sun, J., Liu, H., Zhang, H.: An Extension of Pawlak's Flow Graphs. In: Wang, G.-Y., Peters, J.F., Skowron, A., Yao, Y. (eds.) RSKT 2006. LNCS, vol. 4062, pp. 191–199. Springer, Heidelberg (2006)
34. Suraj, Z., Pancerz, K.: Flow Graphs as a Tool for Mining Prediction Rules of Changes of Components in Temporal Information Systems. In: Yao, J., Lingras, P., Wu, W.-Z., Szczuka, M.S., Cercone, N.J., Ślęzak, D. (eds.) RSKT 2007. LNCS(LNAI), vol. 4481, pp. 468–475. Springer, Heidelberg (2007)
35. Wang, F.H.: On acquiring classification knowledge from noisy data based on Rough Sets. Expert Systems with Applications 29, 49–64 (2005)
36. Wang, Y.: On cognitive informatics. Brain and Mind: A Transdisciplinary Journal of Neuroscience and Neurophilosophy 4(2), 151–167 (2003)
37. Wang, Y.: The Theoretical Framework of Cognitive Informatics. International Journal of Cognitive Informatics and Natural Intelligence 1(1), 1–27 (2007)
38. Yao, Y.Y.: Perspectives of Granular Computing. In: Proceedings of the 2005 IEEE International Conference on Granular Computing, vol. 1, pp. 85–90 (2005)

Author Index